CW00385830

The Poverty of Television

The Poverty of Television

The Mediation of Suffering in Class-Divided Philippines

Jonathan Corpus Ong

ANTHEM PRESS

Anthem Press
An imprint of Wimbledon Publishing Company
www.anthempress.com

This edition first published in UK and USA 2015
by ANTHEM PRESS
75–76 Blackfriars Road, London SE1 8HA, UK
or PO Box 9779, London SW19 7ZG, UK
and
244 Madison Ave #116, New York, NY 10016, USA

British Library Cataloguing-in-Publication Data
A catalogue record for this book is available from the British Library.

Library of Congress Cataloging-in-Publication Data
Ong, Jonathan Corpus.
The poverty of television : the mediation of suffering in class-divided Philippines/
Jonathan Corpus Ong.
pages cm. – (Anthem global media and communication studies)
Includes bibliographical references.
ISBN 978-1-78308-406-7 (hard back : alk. paper)
1. Television viewers–Philippines–Attitudes. 2. Television programs–Social aspects–
Philippines. 3. Suffering on television. 4. Television–Psychological aspects–Philippines.
I. Title.
PN1992.3.P5O54 2015
302.23'4509599–dc23
2015004782

ISBN-13 978 1 78308 406 7 (Hbk)
ISBN-10 1 78308 406 5 (Hbk)

CONTENTS

List of Figures and Tables   vii

Acknowledgements   ix

Introduction: The Poverty of Television   1

1. The Moral Turn: From First Principles to Lay Moralities   15

2. Theorizing Mediated Suffering: Ethics of Media Texts,
    Audiences and Ecologies   39

3. Audience Ethics: Mediating Suffering in Everyday Life   61

4. Entertainment: Playing with Pity   89

5. News: Recognizing Calls to Action   119

    Conclusions: Mediating Suffering, Dividing Class   153

Appendix   177

Notes   187

References   193

Index   211

LIST OF FIGURES AND TABLES

Figure 1. A fisherman on Bantayan Island stands beside his new
fishing boat donated by GMA Network in the aftermath
of Typhoon Haiyan. Photo by the author.   58

Figure 2. Students and alumni of Ateneo de Manila gather in
their sports centre to pack food and other relief goods
for affected communities of Typhoon Ondoy.
Photo by the author.   146

Table 1. Q1 2009 Audience Shares by Channel   55
Table 2. Mapping Moral Responses to *Wowowee*   114
Table 3. Mapping Moral Responses to Suffering in the News 148
Table 4. Respondent Characteristics (Total: 92)   179

ACKNOWLEDGEMENTS

The idea behind this book began during my MSc degree at the LSE, where I studied under the most passionate professors thinking through pressing questions about media morality. Sonia Livingstone, Youna Kim and Terhi Rantanen were among my encouraging mentors. I have to specially thank Shani Orgad, whose lectures seduced me into academia, and introduced me to my eventual PhD supervisor.

This book would not have been possible without Mirca Madianou. Mirca guided me for four years as my supervisor at the University of Cambridge. While Cambridge was an odd place to study television and popular culture (not to mention suffering), Mirca's intellectual navigation and fierce friendship found me a home and shaped my voice as a writer. She continues to teach me to this day – now as colleagues in the Humanitarian Technologies Project – and by being her own example of humility and empathy as 'first principles' of research.

Many ideas in the dissertation are the product of discussions with generous and critical mentors. I am deeply thankful to Lilie Chouliaraki–first, for inspiring me through her lyrical and imaginative writing, and then for always challenging me to write and do better. Danny Miller visited Manila during my fieldwork taught me to be less cautious and think/live more dangerously. Georgina Born's suggestions at a crucial stage of my project informed its eventual focus. Nick Couldry gave terrific advice during my PhD examination.

During fieldwork, I am thankful for the hospitality extended by ABS-CBN and GMA. Had it not been for materials and insights shared by Tina Monzon-Palma, Lita Montifar, Cookie Bartolome, Ryan Chua, Ayee Macaraig, Mel Tiangco, Jobart Bartolome, Ella Evangelista-Martelino, Angeli Atienza, Cel Amores, and Regie Bautista, important characters and subplots would have gone missing from the story. Melo Esguerra, Bryant Macale, Bea Ledesma, and Roland Tolentino clarified many important points about the Philippine media landscape.

For 16 years now, Jason Cabanes has been like a brother to me. Thank you for always getting me on the comeback trail. I would never have considered getting a PhD without Nicole Curato's encouragement–what Nicole says or subtly implies, we should always do. Violet Valdez is my first and remains the staunchest academic mentor. I am very lucky to have a most intuitive 'lay expert' friend in Deepa Paul-Plazo, whose proofread was loving, meticulous, and challenging. Pamela Combinido helped and cheered me through the final stages of manuscript preparation. I am grateful for conversations with peers about my work: Megha Amrith, Kristel Acedera, Jayeel Cornelio, Leloy Claudio, Ranjana Das, Helena Patzer, Guanming Low, Lawrence Santiago, Dominic Yeo, Ozan Asik, and Casey Brienza.

My editors and publisher at Anthem Press have been most patient with me through this process. I am thankful to Tej Sood, Brian Stone, and Lori Martinsek for their support. Mano Gonzales' boundless talent and Raymond Ang's visual literacy are to thank for the book cover.

I am also in debt to all my students in Leicester, Hong Kong, Cambridge, and Ateneo de Manila who engaged all my examples on suffering and media ethics, and were first to hear – and challenge – my ideas.

My deepest love and affection go out to my family: my mother Susan, my brother Jeri, and my Ate Liza. My mom's love makes me who I am. It is from her that I first learned the 'lay moralities' of resilience and dignity. My father Antonio and grandmother Lourdes were excited and anxious when I began my PhD; unfortunately they did not live to see me finish it. I dedicate this book to them.

Finally, I am grateful to the many people that I met during my research. I thank them for their hospitality, enthusiasm, and sharp wit. I regret that the succeeding pages, in focusing on conflict and antagonism, could not also record the many occasions of warmth and humour that we shared.

Leicester, UK
23 March 2015

INTRODUCTION:
THE POVERTY OF TELEVISION

The media may have extended reach, but have they extended
understanding?

Roger Silverstone, *Media and Morality*

This book is about how suffering and poverty are represented on television and
how audiences respond to this process of representation. As such, this book
situates itself within the current debates in media ethics, such as the question
of the media's responsibility to 'the other' (Silverstone 2007) and reflections
on the best narrative techniques in media production to move spectators to 'do
something' about faraway tragedies (Chouliaraki 2006; 2013). But unlike most
work within the moral turn of media and communications studies in recent
years, I approach this set of normative and often philosophical questions from
an audience-centred perspective that puts at the heart of analysis the voices of
ordinary people who engage with television in their everyday lives.

How and when are people moved by images of suffering on television? In
what occasions do they turn away? How do people's social and cultural contexts
shape their responses to news about disaster or to the mediated appearance of
poor people in genres of entertainment television? And how do poor people
and marginalized communities themselves regard the television narratives
that journalists and television producers create about them? Drawing from a
twenty-month ethnography of television programmes and their audiences in the
Philippines conducted between 2009 and 2011, I provide in this book an account
of people's 'actually existing' responses to mediated suffering that have so far been
lacking in recent discussions of humanitarianism, social suffering and media ethics.

Aside from using ethnography to nuance our understanding of media's
purported impact on our compassion fatigue and record observations as to
how social factors of class, ethnicity and age shape people's interest in or
apathy with faraway tragedy, this book retells the story of television as a central

institution in which suffering and disaster are represented and resolved and which, to an extent, is also able to make class by reproducing and amplifying class divides in the Filipino society.

As I show in succeeding chapters, Filipino television is a site in which suffering is spectacularly displayed, rather than sanitized or ignored as in other developing countries such as India (Mankekar 1999; Sainath 2009). In the Philippines, television over-represents the poor across genres of news, game show, talk show and documentary, given that the lower class makes up 74 per cent of the population (Virola et al. 2013) and remains the most lucrative target market of the top advertisers, as their low purchasing power is made up for by sheer numbers. Politically, they are also an important constituency as they are recognized by political elites as *the masa* (masses) that determine elections – contests where television has played the role of 'kingmaker' (Coronel 1999).

More uniquely, Filipino television is also a television of intervention (Ong 2015a), especially during disasters. Not only does Filipino television mediate in symbolic terms, such as in the circulation of representations of the Filipino poor, they mediate in material terms as well. Privately owned television networks dispense aid and assistance not only in wealth-sharing game shows directed to the masses but also in charitable projects and disaster relief operations. During disasters, journalists do not simply produce stories and recruit case studies for their programs, but directly dole out food and clothes to affected peoples. During fieldwork for this book in September 2009, when Typhoon Ondoy/Ketsana hit metropolitan Manila and submerged 80 per cent of the nation's capital, journalists and celebrities used their own helicopters and speedboats to rescue stranded victims and provide them with food; they subsequently broadcast these stories of heroic rescues and benevolent donations on television. In the words of writer Conrado de Quiros,

> In fact the monumental thing that happened was the complete absence of government. The only government there was were the media, notably [privately-owned television networks] ABS-CBN and GMA-7. You can forgive both for advertising their wares, or relief efforts, under the extenuating circumstances. They were the government. They were the central authority apprising the public of the situation. They were the central authority coming to the aid of victims. They were running the country (De Quiros 2009a).

In this light, I intend for the book's title, *The Poverty of Television*, to relate here not only to the ethical challenges of representing suffering and underprivileged people on television in the traditional sense of circulating narratives about the other. Poverty and suffering are treated in this book not only as media content, but as social conditions, material realities and embodied individual selves that television interacts with through their diverse modes of professional practice: storytelling, reporting, interviewing, fundraising, rescuing and giving aid. Poverty *of* Filipino television refers specifically to the idea that poverty is not an 'out there' phenomenon but 'in here' in the specific context of a developing and disaster-prone Philippines; the everyday experience of poverty by those with the greatest intimacy with television (both as content and also as institution) informs the political economy of mass media institutions, the aesthetics of their productions and the process of direct interactions between media people and ordinary people.

The title *The Poverty of Television* is also meant as an invitation towards moral reflection about diverse forms of media practice rather than a pessimistic polemic about a failure of television as a medium for moral education and convivial connection. Television, as with any communications technology, is plural in its possibilities, unpredictable in its consequences and diverse both its institutional infrastructure and technological affordances, such that it can, on occasion, succeed in inviting sympathy just as it can in other contexts facilitate social denial about others' suffering. In this book, I retell stories of media audiences and media producers with the aim of deepening our understanding, identifying the very specific and local consequences of good or bad media practice and facilitating dialogue about media ethics that aims to improve the media's status quo whilst recognizing the value that ordinary people place in the media's peculiar practices of over-representing suffering. While the title echoes EP Thompson's similarly titled *Poverty of Theory* (1978), readers will find that the aim and style here are not those of a polemic but of a conversation that discovers within everyday experience the norms and values that people achieve, fail and aspire to in conditions of great contingency and vulnerability.

A Moral Turn in Media Studies

In the past two decades, various disciplines in the social sciences and humanities have indeed been engaged with normative questions about care and obligation for different others (not just suffering others, but others based on nationality, ethnicity, sexuality, class, religion, disability etc.). Questions concerning this have been raised in sociology (Bauman 2001; Boltanski 1999; Cohen 2001; Tester 2001), anthropology (Kleinman et al. 1997; Heintz 2009; Howell 1997; Laidlaw 2002), social psychology (Seu 2003; Dalton et al. 2008), philosophy (Arneson 2004; Butler 2004, 2009; Singer 2009; Sontag 2003), geography (Cutchin 2002; Smith 1998) and media studies (Chouliaraki 2006, 2010, 2013; Couldry 2006, 2008a; Couldry, Madianou and Pinchevski 2013; Ellis 2009; Frosh and Pinchevski, 2009; Orgad 2008; Orgad and Seu, 2014; Peters 2005; Silverstone 2007).

In media studies particularly, the literature is situated in the Western context, with the assumption that a privileged self is in the position of making sense of faraway people's suffering, usually understood as 'famine, war, death and pestilence' (Moeller 1999, 1). Luc Boltanski in *Distant Suffering* (1999, xv), for one, asks what form can commitment take when those called upon to act are 'thousands of miles away from the persons suffering, comfortably installed in front of the television set in the shelter of their homes'? And in *States of Denial*, Stan Cohen (2001, 169) describes the contradictory position of the privileged Western self: 'On the one hand, immediacy breaks down the older barriers to knowledge and compassion, the TV news becoming a "hopeful example of the internationalization of conscience". But, on the other, its selectivity, promiscuity and short attention time span make viewers into "voyeurs of the suffering of others, tourists amidst their landscapes of anguish."' Whilst there is some research on non-Western peoples' experiences of pain and suffering (Das 1995; Melhuus 1997), including some rich historical and anthropological work on suffering, disaster and idioms of pity in the Philippines (Bankoff 2003; Cannell 1999; Hollnsteiner 1973; Ileto 1979; Jocano 1975), these studies are focused on how vulnerable peoples find ways of coping in light of traumatic disasters or mundane poverty. None of these studies specifically explores people's responses to televised suffering and ordinary poverty and examines how they may likewise reflect on their obligations to 'other' sufferers beyond family or community.

Within the literature on mediated suffering, there have been contentious debates about two interrelated issues: first, the representation of suffering by media producers, and second, the reception of suffering by media audiences.

First, there is a concern about how media producers might create routinized, sanitized, or dehumanized stories of suffering. Studies on the news critique how the genre 'objectifies those it seeks to represent' (Donnar 2009, 1) or selectively privileges certain kinds of suffering over others (Moeller 1999). The genres of talk show and reality television have also received inquiry into how they encourage, or exploit, ordinary people to voice their own suffering in the public sphere (Illouz 2003; Lunt and Stenner 2005; Moorti 1998; Ouellette 2004; Wood and Skeggs 2009). A significant debate about the ethics of representation is whether producers should represent sufferers as humane, empowered and identifiable or should portray suffering at its worst in order to mobilize public action. Some argue that humanizing sufferers is more effective in enabling audiences to effectively relate with those suffering and perhaps even be moved to humanitarian action (Tester 2001), whilst others contend that underscoring the gruesomeness of a disaster leads to more attention and indignant political action (Cohen 2001, 183; Orgad 2008). Many of these studies are text-focused, with a reliance on discourse analysis (Chouliaraki 2006; Vestergaard 2009), content analysis (Orgad 2008) and general impressionistic analysis of media texts (Illouz 2003; Moeller 1999). Some studies meanwhile take a philosophical approach and propose normative prescriptions as to what is the most morally desirable mode of representing suffering (Butler 2004, 2009; Silverstone 2007). Some of the most thoughtful elaborations on representing suffering have since veered away from making inferences as to how audiences at home will respond to media images but recognize texts as 'performative' in how they instantiate social and moral relationships between Western audiences and their distant others without necessarily determining outcomes of reception (Chouliaraki 2012). Whatever the approach, however, the lack of empirical audience research in studies of mediated suffering has resulted in normative accounts of media's and audiences' responsibility that takes the privileged Westerner as the default viewer, with less reflection on how differences of social and cultural contexts shape processes of moral reflection and interpretation in the experience of media reception.

Second, and closely linked to the first, there has been a concern with how television audiences actually receive these representations – whether or not they act to address others' suffering, and whether or not they express compassion or experience compassion fatigue. Using different methods, from surveys (Höijer 2004; Kinnick et al. 1996) to focus groups (Dalton et al. 2008; Kyriakidou 2005; Seu 2003), researchers have described the ways in which (Western) audiences negotiate these demands. They investigate how people avoid uncomfortable media images of pain and suffering or, in other contexts, find points of identification from their own experiences of suffering with the suffering of others in the media. There is an emphasis here on audiences' emotional responses of pity or compassion (Boltanski 1999; Höijer 2004). Whilst some of these studies do employ actual tools of audience research and attempt to account for different responses, the choice of experimental and survey methods still fails to capture naturally occurring responses to televised suffering. These studies are also conducted in Western contexts, and whilst some provide the perspective of people in lower socio-economic classes who have personal experiences of suffering themselves (Kyriakidou 2005), audience studies have yet to record the voices of people whose experience of suffering is not defined by distance but proximity. What do people who self-identify as sufferers themselves have to say about how television represents them? And how might a culture of everyday suffering shed light on the wider debates about compassion fatigue and media ethics?

This book, by insisting on an ethnographic and audience-centred approach to the issue of mediated suffering and by being situated in a context of everyday suffering rather than distant suffering, seeks to contribute to the gaps currently present in these two sets of debates. It aims to provide empirical evidence to nuance and challenge some theoretical assumptions and moral positions currently made in the media ethics literature. It seeks to explore everyday contexts of media engagement as spaces for moral action and reflection, insofar as beliefs and sentiments about strangers and vulnerable others are provoked in media reception. Crucially, it reframes current debates about the moral implications of witnessing distant suffering in [Western] media with a consideration of how suffering is witnessed and acted upon by people who live within everyday contexts of poverty – and their everyday mediations through their televisions at home.

The Context of This Book

As mentioned, this book is based on ethnographic work conducted in Manila, the capital city of the Philippines. The Philippines provides a poignant case to dialogue with current debates on the ethics of mediated suffering, as themes of suffering, victimhood, resistance and resilience have long been evident in canonic works in its anthropology and history (Constantino 1985; Ileto 1979; Jocano 1975; Kerkvliet 1990; Rafael 1988). Modern Philippine history is often retold as a series of oppressions by different colonial masters: Spain (1521–1898), the United States (1898–1946) and Japan (1942–1945). Recent historical work has focused on the martial law regime of Ferdinand Marcos (1972–1986), a period of human rights violations, media censorship and crony capitalism (Claudio 2010). Whilst poverty became a significant area of research in Philippine social science in the 1960s and 1970s (David 2001a[1976]), themes of inequality and stratification have been present in the historiography of precolonial feudal Philippines (Aguilar 1998; Scott 1994). And being the third largest Roman Catholic country in the world, Philippine studies have paid significant attention to the role of religion in beliefs and practices of suffering, compassion and mourning (Cannell 1999; Ileto 1979).

Much of social science literature on the Philippines also records 'creative' and 'inventive' ways in which Filipinos, specifically the Filipino poor, behave and survive in conditions of deprivation, hunger and disaster. Challenging the doom and gloom of macro-statistical data on poverty indices, GDP and/or measures of risk, researchers have attested to the 'resilience' of Filipino poor communities across a diverse range of studies: from studies on informal systems of exchange (Hollnsteiner 1973; Jocano 1975) to religious worship of God and messianic figures (Ileto 1979) to strategies of cooperation and coercion within and across classes (Cannell 1999; Johnson 2010; Kerkvliet 1990; McKay 2009). The work of the geographer Greg Bankoff (2003, 53) even poses a challenge to Western concepts of trauma and hazard in his argument that the Philippines can be viewed as a 'culture of disaster', where 'disasters are not regarded as abnormal situations but as quite the reverse, as a constant feature of life.' He argues against Western depictions of local communities being 'untutored, incapable victims' (17) by documenting their 'historical adaptations' and 'coping mechanisms' that 'accommodate' threat and tragedy (162–70).

Such studies offer useful signposts in thinking about how televised suffering in this context might not be received with Western inflections of 'shock effect' (Chouliaraki 2010, 111–12; Cohen 2001, 203) (as suffering is supposedly regarded in the Philippines as ordinary) or raise questions of doubt (Boltanski 1999: 159–61) (as the existence of suffering could not but be affirmed in everyday encounters or experiences of suffering). However, I also adopt a more critical approach in this book by challenging commonplace assumptions (and celebrations) of the resistance and resilience of the Filipino poor. Following the longstanding challenge of the sociologist Randy David (2001a [1976]), who once called for an engaged Philippine social science that criticizes structural forces that produce inequality instead of celebrating 'coping mechanisms',[1] I intend to reflect on class as a prism by which we can reflect on the ethical problem of suffering upgrade. In the Philippines, 74.7 per cent of the distribution of families by income belongs to the lower class, with 25.2 per cent middle class and 0.1 per cent upper class (Virola et al. 2013). Despite modest increases in the country's GDP and remarkable international investment grades (Martin 2014), the number of families living below the poverty line has remained 'practically unchanged' in recent years (NSCB 2013), and income inequality has been recorded to be the highest among its Southeast Asian neighbours (Ho 2011). Annual reports indicate that in 2012 27.9 per cent of the population lived below the poverty line, with over 24 million people subsisting on less than US $1 a day (NSCB 2013).

With the widening gap between the rich and the poor (NSO 2013), previous studies have documented the elite's patterns of avoiding the lower class in both geographic (Kerkvliet 1990; Tadiar 2004) and symbolic (Ong and Cabañes 2011) spaces, alongside more optimistic narratives of middle-class duty or pity as enabling charitable action towards the poor (Johnson 2010). Exploring compassion and obligation (or their absence) in relation to televised suffering is important, then, given that moral claims to help others are ever present in immediate and mediated appearances of poverty: Philippine media are after all most hospitable to and over-representative of suffering.

As the next chapters will confer, this book additionally seeks to contribute to the nascent field of Philippine media studies. In the Philippines, the prevailing trends in media research remain cultural studies textual analyses

(Tolentino 2001; Devilles 2008; Yapan 2009) and exposés of media organizations (Hofileña 2004; Rimban 1999). Media criticism and popular culture commentary from opinion columnists and cultural elites are often concerned about foreign rather than local productions (De Vera 2009; Dy 2009) because of classed judgments about 'trashy' local media. Media criticism about local television is overwhelmingly driven by concerns over sex and violence and carry classist assumptions about the *masa* (mass) television viewers (ABS-CBN News Online 2010; Cruz 2007; Godinez 2007; Hermitanio 2009). The term *masa* here is often used interchangeably with the poor, who are assumed to lack culture and taste and to be dumbed down by television (Dancel 2010; Stuart Santiago 2011). This book challenges such trends and argues that ethnographic research on the media in everyday life can intervene in public discourse by contributing a more sensitive and nuanced perspective about Philippine television and their audiences. It also aims to carefully show how differences in responses to televised suffering are expressive of classed moralities about suffering and its representation, rather than simply being a matter of personal taste. It exposes how the media are intertwined with and reproductive of long-existing class differences and inequalities in Philippine society. The book also reflects on Filipino television and their conventions of portraying poverty in comparison with other television systems such as that India, criticized as being 'in denial' of poverty (Mankekar 1999; Sainath 2009).

The Aim of This Book

This book proposes a bottom-up approach in investigating mediated suffering. The initial question that it puts forward is: 'How do audiences in their different contexts respond to suffering on television?'

Through ethnographic observation and interviews with different groups of audiences in Manila supplemented with expert interviews in the industry (elaborated in the Appendix), I specifically explore:

1. the different ways that people consume televised suffering
   - what programmes they watch and how and why they switch off
   - what relationships they have with the media (including their direct experiences of interacting with journalists, celebrities and producers)

2. the different interpretations that people have of particular texts
   - how they express compassion for sufferers in particular programmes
   - what moral judgments of right and wrong they make toward particular sufferers

- what moral judgments of right and wrong they make toward media practices of representing suffering
- the moral discourses that people draw from in responding to televised suffering, that is, how audiences' contexts (class, religion, gender, age) inform moral discourses about sufferers and their representations

In exploring these questions, I provide an ethnographic account of the dynamic and dialectical interaction between audiences and texts in a socio-historical context. It is inspired by and expands on the recent tradition in text-centered work as accounts of situated ethics, or *phronesis*, where specific cases are examined as enactments of ethical discourse (Chouliaraki 2006). While text-centered phronetic work can unpack the historically and culturally specific moral values embedded within media forms, we know less about the processes by which audiences make sense of these moral values and televisual invitations to act on other people's suffering. Meanwhile, philosophical approaches, in providing normative statements of how best to represent or respond to suffering, can be criticized as merely revealing the moral position of the researcher rather than exploring the variety of moral positions held by his or her subjects (Heintz 2009; Fassin 2008; Zigon 2007). In contrast, ethnographic approaches allow for observation of 'the dynamic interaction between abstract ideals and empirical realities' and exploration of how normative theory holds up in actually existing situations (Howell 1997, 4).

The approach taken here is partly inspired by the anthropology of moralities. As I discuss in greater detail in chapter one, the anthropology of moralities examines lay moralities in specific socio-historical contexts. I look into people's moral judgments of right and wrong that underpin expressions of compassion or disgust in the context of viewing televised suffering. This has obvious resonances with the spirit of Aristotelian *phronesis* in the textual analysis of Chouliaraki, where the challenge is to identify the co-articulation of universal and particular perspectives within the practices of mediated suffering. Instead of discourse analysis, though, I use the tools of ethnography to look more closely at how people in different contexts and class positions watch television and what kinds of moral reflection are made possible when viewing news about natural disaster or a reality show where people 'perform' their suffering in confessional interviews. Do people feel compassion for particular kinds of sufferers? Why do upper-class Filipinos seem to switch off from local television? Why do poor Filipinos say that they are fans of programmes where people narrate their painful life experiences, and what comforts might such televised suffering bring? And how are people's responses to suffering shaped by their direct experiences with the media and their evaluation of how the media actually mediate suffering?

In answering these questions, I also aim to reflect on the dynamic, contested and morally ambiguous process of mediation. As I discuss further in chapter two, mediation refers to the ways that the media mediate, 'entering into and shaping the mundane but ubiquitous relations among individuals and between individuals and society' (Livingstone 2009, 7). Such a process raises moral questions because the media not only provide people with tools and technologies to communicate with each other, but through narrative and representation they also contribute to shaping people's 'resources for judgment' and sense of responsibility for the other (Silverstone 2007, 44). How the media mediate suffering has been chronicled in recent years in the context of the Asian tsunami, the South Asian earthquake, 9/11, humanitarian advertising and other areas (Chouliaraki 2010, 2013; Katz and Liebes 2007; Kyriakidou 2005; Orgad 2008; Orgad and Seu, 2014; Vestergaard 2009). What this growing literature signals is a growing anxiety not only about issues of power in light of what mediation does but also about issues of morality. How and in what ways do the media connect or widen the geographical, cultural and moral distances between self and other? As the media play a significant role in the constitution of producers and audiences as moral actors, this ethnographic study aims to reflect on the consequences of mediation, as seen in the moral actions that producers and audiences make about not only their distant others but also and ultimately their own selves.

The Structure of the Book

Chapter one is divided into two main sections. The first section is a critical review of the media ethics literature. This section highlights how this area of research constitutes a significant break from the traditional legalistic framing of what we mean by media ethics in its (1) thrust for global or universal media ethics and its (2) rootedness in various traditions in moral philosophy (e.g., virtue ethics, Levinasian ethics, etc.). This section ends with a critique of the limitations in this literature. I argue here that developing an anthropological ethics of media might be necessary to account for thinking about the ethics of day-to-day media practice, the ethics of media audiences and media ethics beyond Western media systems. The second section of chapter one deals with the literature on suffering, reviewing philosophical as well as anthropological accounts of both (1) the witnessing of others' suffering and (2) the personal and social experience of suffering. This review of the literature highlights on one hand the ethical debates as to the human person's responsibility (if any) to act on his or her neighbour's condition of vulnerability and, on the other hand, the empirical accounts of people's coping mechanisms and adaptation strategies in light of natural disaster, conflict and poverty. The review highlights that these stories of witnessing suffering and directly experiencing suffering often run in parallel rather than in dialogue and argues that it is necessary to think about their interrelationships. While the review of the literature here is global in scope, there are focused, short sections that explicitly relate these discussions to the empirical context of the Philippines.

Chapter two is a more in-depth discussion of media studies debates about the relationship between television and their audiences. It particularly develops the framework of mediation in relation to the media ethics debates. I argue here that the current literature on distant suffering lacks a nuanced account of the relationship between televised representations of suffering and the audiences that encounter these in their everyday lives. Text-centred studies overemphasize how news narratives cause compassion fatigue, while audience-centred studies enumerate audience responses with inadequate references to the textual elements and social factors that shape these responses. This chapter also contrasts a mediational approach to televised suffering with recent theorizations of media witnessing. I argue here that while the witnessing literature has provided a guidepost in thinking about the ethical consequences of showing and seeing suffering in media, it nevertheless obscures the normative from the descriptive and universalizes the experience of the witness it speaks about. To address these gaps and develop a more holistic approach to examine televised suffering, chapter two outlines the significance of mediation theory in accounting for distinct ethical questions that arise from specific 'moments' of mediation and how they altogether inform ethical critique of media practice. Chapter two provides a typology of media ethics debates along the moments of text (textual ethics), production (ecological ethics), and reception (audience ethics). The final section of chapter two moves the discussion from the theoretical debates to the particular context of the Philippines and its peculiar media landscape, with reference to the nature of local media regulation, access and consumption. This section discusses the prevailing business model of Philippine television that actually caters its programming to lower-income audiences who make up the majority of the population and thus constitute the most crucial market for advertisers.

After reviewing conceptual debates in the previous chapters, chapter three presents ethnographic data from interviews with television audiences both in slum communities and middle-class neighbourhoods in dialogue with the media ethics literature accounts of compassion fatigue, witnessing and the responsibility of the audience. This chapter focuses on people's access to and consumption of different media and their genres and narratives of suffering. The key finding here is that there are stark differences in audiences' engagement with televised suffering, whereby middle- and upper-class television audiences tend to switch off Filipino television and instead consume international media, while lower-income audiences affectively engage with the diverse television genres that represent narratives about poor, destitute Filipino people like themselves. Challenging the uniform accounts of compassion fatigue, which states that repeated images of suffering lead to switching off by audiences, this chapter highlights how the decision to switch off is influenced by social factors, particularly of class (and to some extent, gender). This chapter also links the empirical finding of classed and gendered consumption patterns of suffering narratives in Filipino television with the moral debates of witnessing and the responsibility of the audience. I highlight here how the social critique of media may in fact serve as a moral justification for switching off from media (and public issues in general).

From a more general account of television consumption of Filipino audiences of different classes, chapter four presents a specific case study of people's responses to suffering in the genre of factual entertainment television. The particular programme selected here is *Wowowee*, a highly controversial and enormously popular reality/game/talk show where people's personal narrations or performances of suffering are instantly rewarded with cash. The *Wowowee* case came as an 'ethnographic surprise' during fieldwork, as it became a common point of reference in discussions about the (lack of) respectability of the Filipino poor as well as the social critique of the media's exploitation of the poor and their perpetuation of a mendicant culture. In dialogue with the normative question of whether it is better to represent sufferers as humane and identifiable or at their worst, this chapter reveals through ethnographic data that audiences have different interpretations and preferences in relation to the agency or victimhood of sufferers in the media. Using idioms of risk-taking and sacrifice, lower-class audiences read off agency in how they see other poor people queue up for long hours in television centres for a chance to participate in game shows like *Wowowee*, whilst upper-class moralities of dignity and shame underpin affective responses of disgust in the shamelessness of the spectacular suffering in *Wowowee*. Additionally, this chapter moves beyond textual ethics discussions by drawing on the ethnographic data about the actual process by which poor participants queue up and are recruited in reality shows. By highlighting the physical and emotional labour that low-income participants endure in their participation in factual entertainment shows, I argue that current concerns about the exploitation of sufferers in the media ethics literature leave out important ethical questions that arise in the media's procedure and process, outside of textual ethics. Because Philippine media also adopt codes of ethics (i.e., focusing on the regulation of sex and violence) framed in the West, much of everyday media procedure of (1) interacting with vulnerable groups, (2) providing social services and 'acting like the government' in the context of a weak state, and (3) blurring journalism with advocacy and charity fall outside of government regulation and professional discussion of ethics. In the context of *Wowowee*, queuing for the game show led to a stampede that killed 71 people in 2006. Accountability of the incident was, however, evaded or deflected by different parties (the television network, the government and stadium security) in their disagreement about crowd control responsibilities on the ground. This chapter then suggests how the empirical data on *Wowowee* production and reception may form a starting point for the ethical evaluation and regulation of such unorthodox media practices.

Chapter five moves on to the genre of news, which is the traditional focus of debates about the representation and reception of suffering. Following chapters three and four, I demonstrate here class-divergent consumption patterns of news and how they are indicative of different interpretations and evaluations of suffering. In dialogue with the questions of how to best represent suffering and the assumed 'veracity gap' that needs to be resolved in media witnessing (Frosh and Pinchevski, 2009; Peters, 2001), I highlight here that audiences' social class once again becomes an important resource when evaluating the authenticity of other people's suffering and resolving whether to switch off, seek more information, or donate/volunteer in relation to a news narrative. I compare and contrast audiences' responses to news about local suffering and news about distant suffering (particularly, the May 2008 earthquake in Sichuan, China). There are more similarities than differences in how audiences respond to distant suffering, whereby upper- and lower-class audiences presented similar justifications for their non-action towards the Sichuan earthquake with reference to a shared national condition of suffering that demands greater attention than international disasters. This chapter also finds that the news, unlike entertainment, is a moral context that is more likely to provoke moral reflection about obligation and action toward suffering others. Finally, just as in chapter four, this chapter expands on the audience reception data with ethnographic accounts of the economic mediations of suffering that happen in the television networks. Using the case of the September 2008 Typhoon Ondoy/Ketsana in Manila, I evaluate the interventionist practices of media in orchestrating rescue operations, collecting and disseminating donations and engendering trust and cohesion in the context of a weak state unable to manage a natural disaster.

The conclusion returns us to the normative debates outlined at the beginning of the book and revisits them in light of the empirical data of the previous chapters. The book argues for the significance of audience ethnographies in contributing to (1) the media ethics literature, (2) the literature on suffering and (3) the literature on mediation and media power. This chapter argues that an audience-centred approach can more productively identify the different ways in which the witnessing and direct experience of suffering are transformed by the media. Previous speculations about why and when compassion fatigue happens are given empirical grounding, just as media witnessing accounts about how audiences supposedly evaluate the authenticity of sufferers are further nuanced due to the study's attentiveness to the relationship of textual elements and audiences' diverse social positionings in the process of mediation. Some concepts critiqued and nuanced here include authenticity, compassion fatigue and proper distance, as they are discussed here in relation to specificities of audiences' socio-cultural and personal contexts. I assert here that normative concepts and approaches, while prone to simplification and generalization, nevertheless provide a useful guidepost in considering ethical issues. Another issue touched on in the conclusion is the ethical evaluation of the responses of different audiences recorded in the ethnography, such as those of switching off from media (observed primarily among upper-class respondents), affective consumption of media (observed primarily among lower-class audiences), the social critique of media, information seeking and donating/volunteering in relation to mediated suffering. I suggest here that switching off can be evaluated as a highly undesirable response in spite of practices of media over-representation of the lower class. The concluding chapter argues that the symbolic and economic mediation of suffering ultimately reproduces and even amplifies class divides due to its contentious practices of over-representation – practices that require further ethical reflection leading to regulation.

Chapter 1

THE MORAL TURN: FROM FIRST PRINCIPLES TO LAY MORALITIES

To speak of reality becoming a spectacle is a breathtaking provincialism.
It universalizes the viewing habits of a small, educated population living in
the rich part of the world.

Susan Sontag, *Regarding the Pain of Others*

This chapter engages with key concepts and debates in three bodies of
literature that are significant to this study: the media ethics literature, the
anthropology of moralities literature and the suffering literature.

This chapter begins by tracing the moral turn in media scholarship in the
past decade and outlines some of the normative frameworks that scholars
have borrowed and developed from moral philosophy. I argue that this new
literature on media ethics can benefit from a more anthropological approach
that examines lay moralities: judgments of right and wrong that people make
in relation to the media. Whilst this argument is put forth within the media
ethics literature (Born 2008; Zelizer 2008), I propose that the anthropology
of moralities helps us to think more carefully about the relationship between
normative principles and empirical realities as well as how we can examine
everyday practices of television consumption as a moral context.

In the second main section of this chapter, I outline the key assumptions
and debates in the literature on the anthropology of moralities. I reflect on
how this framework, which attempts to describe people's moral frameworks as
situated in specific socio-historical contexts, can be used to dialogue with the
media ethics literature.

Finally, in the third main section of this chapter, I focus on existing
approaches to suffering. I review definitions of suffering from moral philosophy
and from anthropology and sociology. I divide the literature into two groups:
the first approaches the problem of suffering from the perspective of the
witness and the second describes the problem of suffering in the context of

victims or sufferers themselves. It is necessary to review both bodies of work
for my study, as my chosen empirical context locates the problem of suffering
in both perspectives. Unlike existing studies in the media ethics literature, my
research finds that most Filipino audiences identify themselves as poor and
suffering. How they negotiate the competing moral demands of obligation to
suffering others on television as well as their own desires for recognition and
reward is a theme that runs through this book. Before we get there, however, it
is necessary to trace the theoretical roots of my research.

Media Studies and the Moral Turn

Media studies has witnessed quite a dramatic moral turn in recent years. The
writings of Lilie Chouliaraki (2006; 2013), Nick Couldry (2006; 2008a; 2012),
John Durham Peters (1999; 2005) and Roger Silverstone (1999; 2002; 2007),
among others, forcefully argue for the centrality of morality in the project of
media studies. As Silverstone (1999, 142) stresses:

> In so far as the relations between human beings now depend on their
> mediation electronically, and our treatment of each other [...] depends
> on their communication through the same media [...] then we have to
> accept the challenge. If we are to understand in [Isaiah] Berlin's words,
> the 'often violent world in which we live', and our media's role in that
> world, then we are de facto engaged in ethical inquiry.

And the 'we' engaged in ethical inquiry, Couldry (2006, 141) demands,
should not simply be academic scholars, journalists or media owners, but also
'all citizens and all who would be citizens'.

Morality here is concerned not simply about procedural 'codes of ethics' or
the 'moral economy of the household' (Silverstone 1999, 140) or about moral
panics (Drotner 1992); it is instead about:

> first principles [...] the judgment and elucidation of thought and action
> that is oriented towards the other, that defines our relationship to her or
> him in sameness and in otherness, and through our own claims to be a
> moral, human, being are defined [...] Ethics [...] is the application of
> those principles in particular social or historical, personal or professional
> contexts (Silverstone 2007, 7).

Whilst scholars are in disagreement about the correct usage of morality
versus ethics (Couldry 2008a) and some interchangeably use morality and
ethics to pertain to normative evaluations of right and wrong (Zelizer 2008),[1]

what is significant here is Silverstone's notion of first principles in describing this new agenda of media morality. The term first principle here signals that Silverstone, along with some contemporary writers in media ethics, emphasizes a deontological, duty-based, approach to media ethics that seeks to prescribe *a priori* moral obligations that should inform media work in order to facilitate a moral and harmonious relationship between self and the other.

Silverstone, and scholars like Peters (1999) and Pinchevski (2005), draw extensively from the moral philosophy of Emmanuel Levinas, whose ethics of alterity is premised on supererogatory obligations of the self to a radically other Other. Levinas' (1969) philosophy emphasizes the moral as a precondition of social life. Critiquing the Western philosophical tradition of privileging the autonomy of the 'I', in which he uses Heidegger as exemplar, Levinas provocatively argues that the 'I' ultimately exists to be responsible for the Other – a responsibility based not on reciprocity (whether the Other behaves in accordance to expectations of the 'I') but on an asymmetrical and infinite kind of responsibility.

As applied to media and communication ethics, this concern for the other manifests itself in various ways. In Peters' (1999) and Pinchevski's (2005) thought, Levinas' ethics of alterity is used to criticize commonplace notions of the relationship between speaker and receiver. They, for instance, challenge traditional models of communication premised on the speaker being 'perfectly understood' by the listener as a result of the elimination of 'noise' in the technological transmission of messages. For them, the most ethical and genuine encounters with the Other depend in fact on the 'breaks', 'ruptures' and 'interruptions' in processes of communication (Peters 1999, 61–2; Pinchevski 2005, 7).

Such a belief resonates too in Silverstone's *Media and Morality*, described by one of his readers as sounding like an 'epistle' from the force of its normative arguments, rare in media scholarship (Dayan 2007, 113). Drawing again on Levinas, as well as Hannah Arendt, *Media and Morality* (2007) conceptualizes responsibility as the 'first principle'[2] in guiding the work of media producers and their audiences in what he calls the mediapolis: 'the mediated public space where contemporary political life increasingly finds its place, both at national and global levels' (Silverstone 2007, 31). As 'the space where I appear to others as others appear to me' (ibid.), the mediapolis is similar to the normative idea of the public sphere (Habermas 1989). But whereas the public sphere is premised on the exercise of reason in the form of rational-critical discourse, for Silverstone the mediapolis has a fundamentally moral dimension: that of responsibility without reciprocity. Actors in the mediapolis, Silverstone demands, must be concerned with the other: depicting the other, listening to the other, relating to the other and providing the other with the means to respond and represent herself or himself in the common public space of appearance that the media have come to constitute.

The mediapolis, Silverstone also contends, is a cosmopolitan space. Cosmopolitanism here, it must be noted, is both a normative as well as an empirical category for Silverstone. As a cosmopolitan space, the mediapolis is ideally a space welcoming of difference, welcoming of diverse others whose differences are marked by nationality, ethnicity, religion, sexuality, class etc. Cosmopolitanism here harkens back to Enlightenment universalism, which purports that 'every single human being is worthy of equal moral concern and ought to have an allegiance to the community of humankind' (Nowicka and Rovisco 2009, 2). An ethical critique of the media premised on the normative principle of cosmopolitanism stems from how national media systems, individual media texts and domestic audiences might become preoccupied with their interiorities and narcissistic desires whilst ignoring the presence of the other in the global media space.

While the approach to seek out first principles is taken up as well by scholars such as Couldry in his earlier writings (2006) and Dayan (2007), there has also been a move toward a more situated ethics of media. Most influential here is Lilie Chouliaraki, who draws upon Aristotelian *phronesis* to approach mediated suffering where 'each particular case is a unique enactment of ethical discourse which, even though it transcends the case, cannot exist except outside the enactment of cases' (2006, 6). This approach simultaneously attends to the interaction of universals and particulars, where analysis of cases is meant to expose the prevailing moral values in society as instantiated and performed by speech and images. In *The Spectatorship of Suffering*, news reports are assumed as having agency in their capacity to circulate value-driven moral proposals for audiences to act (or not) on distant suffering just as these values are themselves already existing in broader culture.

In both universal (Silverstone 2007) and situated ethics approaches (Chouliaraki 2006; 2013), the exploration of moral values in relation to people's obligation to distant sufferers assumes the standpoint of the Western spectator. The media, in Chouliaraki's words (2006, 83), are implicated in the global divide between 'zones of safety' and 'zones of danger', where media images and texts instantiate moral relationships between Western viewers and their distant suffering others. Her discourse analysis is concerned with the diverse ethical claims made available by news texts, as they variably invite viewers to care for but also at times ignore cases of distant suffering. The other in the case of distant suffering refers to the vulnerable others typically understood to be outside of the Western zone of prosperity, for whom Western media producers and audiences should nevertheless express care but often, in reality, fail to. I elaborate further on this in the next chapter, but for now it is important to note that here Chouliaraki's situated ethics approach is also concerned with normative value judgments. Like Silverstone, Chouliaraki prefers the maximal ethics of cosmopolitanism over communitarianism and sees the former as the desirable moral relationship that should be cultivated by mundane acts of mediation. Her discourse analysis critiques visual and rhetorical elements of news reports that reserve care and identification for sufferers 'like us' (a communitarian principle), preferring reports that involve what she views as a 'morally superior' demand to care for those 'not like us' (a cosmopolitan principle) (84–93).

In any case, this move towards normative media analysis is not without its critics. Zelizer (2008), for one, questions the attraction of media philosophers to universal norms that disregard the historical and geographical contexts of media production and reception. In her historical review of 'about-to-die' photographs in Western newspapers, she argues that moralities are always articulated from particular places and times; therefore, she is unwilling to propose a singular or universal first principle that supports definitive moral judgments on the many photographs that she had collected. Certainly, whether non-Western contexts of media production and reception should be judged according to the same moral principles articulated by the scholars above is a question that remains to be reflected upon in both philosophical writings and empirical research.

Similarly, Born (2008) is skeptical about the philosophical writings on media ethics. As an anthropologist of media, Born leans towards the use of ethnography in recording and analysing the micro-practices of people using and producing media and admonishes philosophers for 'not talking to people'. Born argues for an 'anthropological ethics' of the media that pays close attention to 'alchemical processes' that translate normative philosophical principles to the ethical conduct of media professionals and users, particularly in the context of television organizations. Rather than beginning with articulations of philosophical first principles, Born proposes that ethical critiques should focus on the decisions and practices that media people make and how these are related to their personal interpretations of good and bad media practice.

Indeed, this debate in media ethics can be likened to traditional debates between philosophy and anthropology and debates within strands of moral philosophy itself. In the introduction to the book *Anthropology of Moralities*, Heintz (2009) reviews the historical impasse between universalism and particularism that has made criticism of moral norms and cultural practices difficult. She notes that anthropologists traditionally prefer to describe rather than pass judgment on the moral frameworks of their subjects. Meanwhile, within moral philosophy, scholars are in disagreement about normative frameworks premised on 'moral maximalism' (Slote 2007) that are evident in the works of Levinas (1969) and his media studies interpreters, such as Silverstone (2007), Peters (1999) and Pinchevski (2005). Some propose more modest approaches where moral obligations are confined to a select few rather than generalized and abstract others (Arneson 2004; Tronto 1993).

The tensions between the universal and the particular, between the normative and the empirical, are present throughout this book. Personally, as a sociologist whose disciplinary roots are informed by the humanistic discipline of philosophy, I find the forceful voices advocating a philosophical approach to media morality profoundly significant, providing new analytical lenses by which media criticism (and appreciation!) can be carried out. Media criticism in the academe has traditionally been amoral, as Silverstone (1999, 139–41) has claimed. I have also objected to how local media criticism in the Philippine press is often focused on the censorship of sex and violence, failing to engage with the principles that underpin legal codes and thus being too easily dismissed by defensive, at times even self-righteous, media actors (Ong 2010).

This book situates itself within this literature in media ethics and seeks to continue the discussion of first principles in ethical critiques of media and the tenability of such an approach. At the same time, following Born's (2008) suggestion for an 'anthropological ethics' of media and adopting insights from the productive field of the anthropology of moralities, this study seeks to develop a bottom-up approach so far lacking in the literature. Using the tools of ethnography, this study aims to reflect on how and whether normative principles may hold up through an account of lay moralities of media audiences in specific empirical and historical contexts.

The approach that I take here is not purely relativistic qua Zelizer (2008), who lacks any intention to provide moral judgment on the media productions she studied, nor is it unabashedly universalistic qua Silverstone (2007), whose bias for cosmopolitan moral norms is front and centre. Following the anthropology of moralities, which itself shares the spirit of the situated ethics of Chouliaraki's phronetic analysis, this book intends to explore the interaction between abstract normative ideals within concrete empirical situations. It can also provide judgments of good and bad media (and audience) practice not only in relation to people's violations of philosophical norms, but also in observations of actual consequences of media production and consumption in a particular culture.

Moralities of the Media Audience: Towards an Anthropology of Moralities

This section outlines how the anthropology of moralities makes use of both normative theory as well as empirical methodologies in describing and analysing moral practices and discourses in particular fields and/or local communities.

For philosophers such as David Miller, a perspective from below is said to enable critical reflection as to how philosophical first principles could be properly applied (or not) in everyday life contexts. He says,

> [A]lthough it would be asking too much to require that there should be a spontaneous fit between the claims of the theory and the beliefs that people currently hold, it is important that where there are divergences it should be possible to give people grounds for altering their beliefs in line with the theory (1992, 588).

Miller, a moral philosopher, admits to the limitations of his discipline in how certain philosophical approaches fail to consider the diversity of peoples and their conflicting beliefs and practices within and across cultures. For him, one use of empirical methodology in moral philosophy is to enable further philosophical reflection that might produce better normative prescriptions that people could more realistically emulate and abide by in their everyday lives.

Whilst this book similarly values the role of empirical methodology in academic reflection on ethical issues, it does not primarily seek to test theory in order to produce better normative frameworks, as suggested by Miller above. In this case, I am more sympathetic to the project of the anthropology of moralities in its desire to sketch from the bottom up the moral codes of specific fields or communities, analyse how these codes came to be developed and trace how these might differ from context to context – or even be locally contested by multiple actors (Howell 1997; Heintz 2009; Fassin 2008; Laidlaw 2002; Zigon 2007). Rather than uniformly prescribe a moral framework for all media or all audiences, this book embarks on a grounded analysis of lay moralities or lay normativities in a specific socio-historical context (Howell 1997; Heintz 2009). As Archetti (1997, 100) explains,

> We can analyse morality when we look at explanations of disease and misfortune, of fate, of gender relations, of marriage and family, of accepted games and so on. Action and beliefs are key indicators not only of moral codes but also of contexts and actors. The anthropological analysis allows us to find local moral codes by examining contexts and concrete discourses.

The anthropology of moralities is concerned with contexts that elicit lay discourses of good or bad and right or wrong. In contexts as diverse as Argentinean football (Archetti 1997) and Irish childrearing (Sykes 2009), anthropologists assert that by discussing 'morally problematic scenarios' with their subjects, 'unspoken and unconscious moral values' are given voice in conversation and interaction (Heintz 2009, 12). By provoking people's moral reasoning through 'moral dilemma interviews', supplemented by participant observation, anthropologists can identify local moral codes within which people act or are expected to act.

Further, this sub-field is concerned with the interaction between moral discourses (i.e., enunciations or justifications) and practices (Heintz 2009, 13). By using life story interviews, anthropologists can gather data by asking people to explain values that have guided their life choices. Whilst interviews might elicit 'Sunday-best beliefs', it is understood that through use of participant-observation methods, anthropologists are able to compare their personal narratives with actual practices and posit explanations for both consistencies and discrepancies.

Though research in this area is diverse, a shared thrust is to analyse the 'dynamic interaction between abstract ideals and empirical realities' (Howell 1997, 4). In this way, this field of research engages with the universalism versus cultural relativism debate that has been present not only in discussions between anthropology and moral philosophy but also within the media ethics literature that I reviewed earlier. As Heintz (2009, 5) reviews in *The Anthropology of Moralities*,

> Universalism presupposes the existence of a common core of rationality/ morality from which diversity emerges in response to different natural contexts and as a result of different historical developments [...] Cultural relativism asserts that what we hold to be true/good in one culture can be held to be false/wrong in another culture without any possibility of deciding whether one or the other culture is mistaken in asserting it: each culture has its own rationality. In its strong form, cultural relativism implies that the rationality/morality can only be judged from within a culture and through its own criteria, thus rendering cross-cultural comparison impossible.

Heintz admits that anthropologists studying moralities are themselves conflicted between the two positions, and that scholars sometimes 'ignore the question altogether and even switch unintentionally from one position to the other under the influence of events observed' (6).

In any case, she proposes that the analyst

> has to confront facts and discourses, search for reasons behind the actors' positions (be they 'traditional' or 'modern' or 'Westernized'), measure their engagement in the debate and see how opinions are polarized

within society. [The researcher] cannot surrender to his own emotional and/or moral position, but has to account for the complexity of a phenomenon that reveals which beliefs, values, and meanings underpin action in another society (7).

Laidlaw (2002) proposes a different resolution to the universalism/ cultural relativism dichotomy. For him, it is significant to begin the study of morals with an analysis of the freedom of individuals to choose or not their way of life in a given society. This perspective is useful in that it asks the researcher to examine how individuals are situated in different social positions and work through different affordances and constraints in developing belief systems and translating personal values into action. In the intensely class-divided society where my study is set, Laidlaw's proposal is a challenge to take into account how both elites and sufferers themselves are able to articulate moral positions in responding to televised suffering and how these positions might be shaped by interests, expectations and limitations of their class position.

In summary, the perspective of the anthropology of moralities is useful for my study for three main reasons.

First, I consider media consumption as a dynamic context where lay moralities are reflected upon and expressed. Rather than view the context of media consumption as one of pure enchantment (de Zengotita 2005) or simulation (Baudrillard 1994), I view the act of watching televised suffering as one wherein moral reflection and responsibility are carried out (Silverstone 1999, 2007). By listening in to media talk in relation to televised suffering, I argue that researchers can identify statements of good and bad and idealizations of certain moral values (e.g., authenticity, respectability) in expressions of what they like and do not like about the media. Further, and more specific to this book, lay moralities are not only excavated or captured through media talk; in the process of being interviewed in the context of moral dilemmas, people are provoked into moral reasoning and reflection (Zigon 2007). By probing how people justify their actions (or non-action) toward mediated sufferers and prompting reflection as to whom they consider better or more worthy sufferers, lay moralities move from the realm of abstract ideas to expressed discourse within the research encounter.

Second, the anthropology of moralities is useful for its simultaneous interest in people's discourses as well as their practices. By comparing and triangulating data from interviews and participant observation of people's everyday lives, I aim to go beyond possible utterances of 'Sunday-best beliefs' and analyse the relationship between people's immediate responses to suffering in the media with social and cultural beliefs and practices.

Third, my work shares the intention of the anthropology of moralities to dialogue with pronouncements of normative theory by examining how normative expectations made on audiences play out in everyday life. Indeed, moral philosophers admit that ethnography potentially offers rich 'sociological explanation[s] of some sort as to why beliefs differ so radically' (David Miller 1992, 588). As we will find in later chapters, my fieldwork actually uncovered stark class differences in people's moral discourses and practices. The category of class is important in ethical analyses, since it is argued that the study of ethics begins with a study of people's freedoms and constraints – their access to resources or lack thereof (not only economic, but also styles of thought and modes of reasoning) (Laidlaw 2002; Skeggs 2004; Skeggs et al. 2009). A bottom-up approach enables presentation then not only of the moral standpoint of the researcher or expert but also the multiplicity of moral discourses and practices of ordinary people in relation to the media.

The following section now takes us to the issue of suffering and the various philosophical and empirical approaches that deal with it.

Theoretical Approaches to Suffering

In the media ethics literature, the word 'suffering' is often used to refer to 'famine, death, war and pestilence' (Moeller 1999, 1) that occur in distant places and cultures (Boltanski 1999), specifically in non-Western 'zones of danger' (Chouliaraki 2006, 83). Suffering is usually a catch-all term for events involving pain and death of distant strangers. How and why particular events of tragedy in different Western and non-Western countries are accorded different degrees of attention and visibility – either becoming media events (Dayan and Katz 1992), disaster marathons (Katz and Liebes 2007) or neglected 'orphaned disasters' (Annan in CNN International 2005) – is of significant interest to media ethics scholars.

Whilst such definitional shorthands are useful, it is necessary to review more basic premises about suffering elaborated in related fields. The anthropologist Young (1997) identifies dichotomies between biological and social definitions of suffering. The former focuses on identifying states of somatic pain in organisms, and the latter emphasizes the 'psychological, existential, or spiritual' dimensions of pain. He says that this second kind of suffering has a 'social or moral dimension, in the sense that it is understood locally, by identifiable groups and communities, in the context of ideas about redemption, merit, responsibility, justice, innocence, expiation, etc.' (Young 1997, 245). Similarly, Kleinman et al. (1997, ix) argue that a social approach to suffering is productive in (1) describing linkages between personal problems with wider social problems and (2) analysing cultural specificities in the experience of suffering better than psychological, medical and individualistic approaches.

Given these insights, this book adopts a social rather than an individualistic approach to suffering. As my chosen empirical context includes respondents who self-identify as sufferers themselves, it is necessary to be attentive to local cultural understandings of suffering that people may draw from in their responses to televised suffering. These local codes can then be compared and contrasted with the Western-centric and philosophical discussions in the literature. Within the Filipino context itself, differences may exist in beliefs and practices about suffering that may be shaped by differences in class, religion, gender and age. As I elaborate in this chapter, previous empirical studies attest to the significance of these sociological factors in influencing how people help suffering others or cope with suffering themselves.

In the succeeding sections of this chapter, I review studies that discuss what I consider two rather disparate but fundamentally interrelated strands of thought that elaborate on the moralities of suffering. The first problem is what I call the 'problem of the witness'; this section reflects on philosophical and theological debates about the moral obligations of a witness, or bystander, to the suffering other. Included here is a section on Philippine scholarship about compassion and charity that is relevant to the 'problem of the witness'. The second problem is what I call the 'problem of the sufferer'; this group of studies, mostly from anthropology and sociology, describes how communities coalesce out of common experiences of tragedy and how they manage to cope in conditions of suffering. Also included here is a discussion of Philippine social science scholarship on suffering, historically described and criticized as a 'sociology of coping mechanisms' (David 2001a[1976], 42). These sections about suffering in the Philippines also begin the trail of the book's discussion of the poverty of television, as we expand on the media ethics literature on witnessing distant suffering and explore the lower-class majority's everyday engagements with television content and momentous interactions with media producers.

Suffering and the problem of the witness

Before thinkers such as Boltanski and Silverstone elaborated on the moral problem of knowing about distant suffering as a result of media, in Western thought the French philosopher Voltaire reflected on the news about the devastating earthquake that hit Lisbon in the year 1755. When the news of the disaster reached the French philosophers, it sparked 'one of the most famous philosophical and theological controversies of French intellectual history' (Illouz 2003, 189). In his *Poéme sur le désastre de Lisbonne*, a distraught Voltaire is recorded: 'Which crime, which mistake have these children committed, crushed on their mother's breast in their own blood? Was Lisbon, which is no more, more corrupt than London or Paris full of delights? What? Lisbon is destroyed and we dance in Paris?' According to Illouz (2003, 189–90), Voltaire's intervention is significant in that it marks the first time that a philosopher directly addressed his community of fellow philosophers and the general public about a present-day but distant disaster.

Indeed, prior to philosophical reflection about humans' obligations to distant others, it was religious theodicy that developed the intellectual history of the ethics of suffering. Charity, after all, is 'the great virtue' of the Christian West (Comte-Sponville 2001, 289). Philosophers and theologians have also noted the close linkages between charity and its sister virtue, compassion. According to Comte-Sponville, compassion is in fact 'the principal content of charity, its truest affect, indeed its real name' (289). Charity here must be understood not simply as the practice of almsgiving but as love. Charity, derived from the Latin *caritas* and the Greek *agape*, pertains to a 'universal love, without preference or choice, a dilection without predilection, a love without limits and even devoid of egoistical or affective justifications' (284).

Meanwhile, compassion is a more modest and reactive form of charity as it 'requires the other person's suffering in order to love; it needs its cripple dressed in rags, depends on the spectacle of his misery' (Jankelevitch 1986, 168–9). In sociology, compassion is defined as the virtue that 'involves an active moral demand to address others' suffering' (Sznaider 1998, 117). In moral philosophy, compassion is defined as sympathy in pain or sadness or, in other words, a participation in the suffering of others (Scheler 1970, 5). It is the opposite of cruelty, which rejoices in the suffering of others, and of egoism, which is indifferent to others' suffering (Comte-Sponville 2001, 106).

Compassion has received much attention from scholars in various disciplines in reflecting on the question of the scope of (presumably Western) individuals' care and responsibility for the suffering of nearby strangers and distant others. However, it is not without its critics.

The first main criticism of compassion involves its reactive character. The sociologist Tester (2001, 65) is skeptical about the ethical potential of compassion as it contains a 'jack-in-the-box quality', wholly dependent on external stimuli for the virtue to enact itself. In this sense, compassion is not self-sustaining, as it can be lost in the absence of suffering, or, as media studies in the next chapter suggest, compassion can make way for compassion fatigue, which involves an active avoidance of abundant and disturbing images of suffering (Cohen 2001, 187–92).

The second main criticism of compassion focuses on its projective quality, specifically its way of identifying with its object. The political philosophy of

Arendt (1963, 85) is illustrative of this: compassion, or in her words 'pity', she says, 'has proved to possess a greater capacity for cruelty than cruelty itself'. In her book *On Revolution* (1963), Arendt warns against what she calls the politics of pity, that is, a politics based on compassion for the suffering of another person. Arendt prefers action to be carried out between equal partners in human dignity. She calls for justice rather than sympathy and for principle rather than emotion. In another text, Arendt (1968) argues that compassion reduces all distance between people. In Illouz's (2003, 203) interpretation, Arendt is wary about how compassion abolishes what she calls the 'in-between' precisely because compassion is based on an immediate identification with the sufferer. Arendt's writings, it must be noted, inspired those of Silverstone (2007), as I will discuss in the succeeding chapter.

The third main criticism of compassion concerns its place-bound quality, specifically its predilection for proximal over distant sufferers. Ginzburg (1994, 108) notes that in his *Rhetoric*, Aristotle says that 'the nearness of the terrible makes men [sic] pity. Men also pity those who resemble them [...] for all such relations make a man more likely to think that their misfortune may befall him as well.' For some moral philosophers, particularly those espousing communitarianism, it is understandable that people would demonstrate compassion for the nearby neighbour or family member. As Etzioni (1995, 146–7) suggests, 'We start with our responsibility to ourselves and to members of our community; we expand the reach of our moral claims and duties from there.' Certain strands of feminism also think favourably about caring for a particular other rather than abstract others. As Clement (1996, 16–17) notes, feminist writings emphasize the importance of knowing the concrete other for whom we care, in the active sense of caring for rather than caring *about*. Noddings even provocatively argues that caring *for* distant peoples is 'care in name only: we cannot care for people we do not know' (in Smith 1998, 29).

Not all scholars, however, focus on compassion in developing their own ethical frameworks. For instance, Singer's moral philosophy, directed primarily to people in affluent nations as regards their responsibilities to individuals in poorer parts of the world, elaborates, 'If it is in our power to prevent something bad from happening, without thereby sacrificing anything of comparable moral importance, we ought, morally to do it' (Singer 1973, 23). This principle takes no account of distance: 'If we accept any principle of impartiality, universalizability, equality, or whatever, we cannot discriminate against someone merely because he is far away from us' (24). Singer's philosophy is not based upon emotion but on reasonable reflection, and is described as 'supererogatory' in equating what is morally right with what is morally obligatory (Arneson 2004, 51–6). In many ways, Singer's thought is similar to the moral maximalism of media ethics scholars such as Silverstone (2007). We should also note that his proposal stresses the urgency of proper redistribution (Fraser 1997; Fraser and Honneth 2003) of material wealth as a moral obligation for individuals living in relative comfort compared to the world's poor (Singer 2004).

The problem of the witness in the Philippines

In a landmark review of the field, the sociologist David (2001[1982], 31) traces the development of Philippine social science as traditionally interested in identifying and describing Filipino moral norms and values. It is perhaps unsurprising that across canonic work in different disciplines, discussions of moral codes underpinned by *damay* (mourning/sympathy), *awa* (compassion/ pity), *mauulung* (pity), *malasakit* (sympathy/sorrow), *hiya* (shame), *pakikisama* (getting along smoothly) and *utang na loob* (debt of gratitude) repeatedly appear. Moral codes of how to act in relation to others, including suffering or less privileged others, are discussed across a wide variety of historical and anthropological studies (Aguilar 1998; Cannell 1999; Hollnsteiner 1973; Ileto 1979; Jocano 1975, 1997; Johnson 2010; Lynch 1973; McKay 2009; Pinches 1999; Rafael 1988).

It is significant to note that the constellation of concepts above implies a social order premised on relational rather than individualistic moral codes. A classic text that aimed to explain the country to foreigners, *Culture Shock! Philippines* (Roces and Roces 1992) used the metaphor of fried eggs to contrast Filipino and Western selves. Westerners are individual fried eggs whose edges do not touch; Filipinos are eggs fried together so that their whites blend, leaving a pattern of yolks embedded in a wider field. This metaphor resonates with Jocano's (1997, 61) explanation of the moral norm *kapwa* – the 'consciousness of reciprocally shared identities' – that underpins social interactions. *Kapwa* compels Filipinos to be consistently attendant of others' feelings. It places a premium on conflict avoidance and places a negative value on having a 'mind-your-own-business' or 'each-to-his-own' attitude (66–7). According to Jocano, *kapwa* also resonates with the Catholic[3] belief that human persons live as equals in fellowship with each other by virtue of all being children of God (63).

The moral standard of *kapwa* is indeed still profoundly evident in contemporary ethnographies, such as McKay's (2009) work with transnational families. McKay found that labour migrants' practices of sending remittances and gift boxes to family members, extended kin and the community left behind are perceived as expressions of a moral obligation to others rather than mere acts of conspicuous consumption. McKay expands on Strathern's

(1996) notion of 'cutting the network' and describes how network-cutting practices of avoiding contact and ignoring requests for economic assistance are considered violations of relational norms. Those who violate such norms are labelled as having no shame (*walang hiya*) and no debt of gratitude (*walang utang na loob*) – some of the most morally damaging statements one can make to a Filipino (see also Miller and Madianou 2010).

Another contemporary ethnography that reflects on moral obligations, this time of middle-class migrants towards lower-class labour migrants, is Johnson's (2010) work in Saudi Arabia. Middle-class migrants here are IT professionals and engineers serving Saudi Arabia's booming economy, whilst labour migrants are female domestic workers and male construction workers. He observes that middle-class migrants leave their homeland with their entire families in tow, as their salaries could provide them a comfortable lifestyle in the new country. Labour migrants, meanwhile, travel alone and work in order to send remittances to children and relatives left behind. Johnson observes that the practice of (female) domestic workers leaving behind their children is judged by the middle class as irresponsible parenting, with middle-class migrants labelling them 'bad girls' (443). However, whenever domestic workers run into crises, such as physical abuse from their Saudi employers or the termination of their work contract, middle-class migrants respond with *mauulung (pity)* and provide them protection by employing them in their homes. Whilst ordinarily these two migrant groups rarely interact and are suspicious of each other, during moments of crisis the boundaries collapse as a function of *mauulung* for the other's misfortunes. Johnson's ethnography skillfully demonstrates competing and contradictory classed moralities at work in the Filipino diaspora and certainly challenges the more functionalist and essentialist accounts of Jocano (1997), which depicts a singular and totalizing framework of a Filipino morality.

Class politics is indeed a common theme in the literature on Filipino social relations. Kerkvliet (1990, 14) writes about the everyday politics of peasants and landlords in the northern Philippines, continually 'trying to make claims on each other'. He found that poor people try to attach themselves to those with more resources to protect themselves against adversities and improve their own claims to resources. Whilst their relations may outwardly appear as smooth and harmonious, his ethnography uncovered the 'tension, antagonism, and conflict' expressed in contending moralities of justice and fairness in relation to the distribution of resources (16–17). Whilst Kerkvliet (1990) and others (Aguilar 1998; Cannell 1999; Ileto 1979; Pinches 1987) concede that the poor exhibit creativity and agency in the ways that they make claims for pity – and accompanying

economic or political assistance – from privileged patrons, some studies
suggest that the upper classes in fact resist or ignore such claims when
these conflict with their class interests or contradict their moralities
of respectability, hard work and independence. Kerkvliet (1990, 168)
finds that one strategy landlords used to ignore labourers' claims was
geographical distancing: 'by distancing themselves from as many of the
poor as possible, better-off people reduce their obligations'. In Tadiar's
(2004, 85–92) writings about Manila – the field site of this study – she also
reflects on the urban fixture of the flyover as expressive of geographical
and social boundary work between classes. She argues that flyovers, which
are elevated motorways along the metropolis, enable the upper strata to
literally transcend the sites of squalor and poverty that exist beneath them.
Her thesis of 'fantasy-production' hints at a kind of social denial that
Filipino elites engage in to 'fly over' the squalor of poverty and ignore the
presence of less privileged others in urban life. In Tadiar's words, 'flyovers
perform the function of corridors leading to and from the exclusive,
walled-in neighbourhoods where the upper strata are ensconced' (91) and
'[f]rom this suspended pathway the city looks greener because the foliage
of walled-in neighbourhoods become visible, and the roofs of shanties
look like variegated pieces of mosaic or a collage' (84).

Be that as it may, natural disasters and personal crises are documented in
the literature as extraordinary events during which class, family, or personal
interests are set aside (Bankoff 2003; Jocano 1997; Johnson 2010). Jocano
(1997, 68) attests to the moral expectation that 'one has to go out of his [sic]
way to condole, sympathize, or share in the sorrows of others' during crises;
even enemies are expected to reconcile.' Even the poor, though they have
few resources to assist victims of calamity, are expected to at least 'pray for
others' or 'do good deeds' (Atillo and Serrano 2004, 126). Religious beliefs
and practices are said to inform acts of giving as '[Christian] believers are
socialized to the idea that any act of generosity done on earth is a ticket to
heaven' (ibid.). In addition, my study on Filipino migrants reveals that elite
and middle-class Filipinos internalize a deep sense of obligation to give back
to their poor country and view their return to the homeland as a fulfillment
of this obligation: 'by coming home, "they are able to directly transmute
through their bodies the ideas, talents and skills absent in [what they believe
is] a deprived and depleted Philippine society"' (Ong and Cabanes 2010, 219).

This book is crucially interested in the ways in which elite and middle-
class Filipinos engage with mediated distant and proximal suffering between
these dialectical tensions of on the one hand 'flying over' the poor in public
life and on the other hand dealing with culturally embedded imperatives of
compassion and duty toward less privileged others.

Suffering and the problem of the sufferer

When suffering is discussed from the perspective of the sufferer instead of the witness, the first significant theme pertains to ways in which people cope with experiences of pain and hardship. As mentioned earlier, religious thought and tradition have provided a rich backdrop against which suffering is understood and experienced by believers.

For much of the Christian West, the New Testament has been a key source on how to conduct oneself in light of various life difficulties:

Blessed are the meek: for they shall inherit the earth (Matthew 5:5)

It is easier for a camel to go through the eye of a needle, than for a rich man to enter the kingdom of God (Mark 10:25)

For philosophers such as Nietzsche (1998[1889]), Christianity is to be criticized for providing consolation to sufferers, as it weakens the resolve to take responsibility for personal faults and social problems by conferring power to environmental or divine forces. In spite of such criticisms, philosophers such as Schopenhauer (2007[1840]) are more sympathetic with the idea that undergoing experiences of suffering will produce insight and enlightenment for the sufferer who properly reflects on her or his condition.

It is important to note that some of the most interesting discussions of how sufferers cope in experiences of pain and difficulty can be found not only in philosophical writings but also in anthropology, sociology and cultural geography. Empirical studies that provide historicized and contextualized accounts of people's experiences of suffering are then useful to review.

In the anthropology of moralities literature, the work of Melhuus (1997) on the meanings of suffering in Mexico reveals both gendered and religious understandings of suffering. In Christian-Catholic Mexico, he found that women were more likely than men to voice commentaries about life situations of suffering, though strictly following religious (and local) understandings of sexuality and motherhood. For Mexican women, patient endurance of suffering is their expected moral task, whilst Mexican men are responsible for defending the women's honour against those who might threaten the women's dual status as either innocent virgins or dutiful mothers (187).

Melhuus' insights about the ways in which female sufferers are authorized to speak about their suffering according to particular norms are also reflected in more formal contexts, such as legal restorative justice sessions that Kenney and Clairmont's (2009) ethnography describes. In their work on youth restorative justice sessions in Canada, Kenney and Clairmont argue that adopting the role of the victim is strategically deployed as 'both sword and shield'. Through rhetorical strategies and displays of emotion, people engage in 'victim contests' in the courtroom so as to either claim damages ('victim as sword') or avoid prosecution ('victim as shield') for whatever pain or suffering they had experienced (303).

Certainly, these studies relate back to the idea of everyday life as a terrain of power struggle and resistance. The landmark ethnographic study of James Scott (1985) in the context of working-class peasants in Malaysia is instructive here, as he coins the term 'weapons of the weak' to pertain to subaltern soft forms of resistance (foot-dragging, sabotage, feigning ignorance) that marginalized groups practise in the face of threatening hegemonic forces.

It must be noted, however, that performances of victimhood or resistance do not always have clear and specific objectives when carried out. Whilst Kenney and Clairmont (2009) and Scott (1985) may hint at political and material objectives in their subjects' desire for legal retribution or class emancipation, other anthropologists describe more symbolic qualities associated with performances of victimhood and mourning. Das (1995), in her ethnographic work with Punjabi Hindu survivors of the Partition riots in India, found that female victims mourned and grieved over their lost family members in practices confined to their homes and local community. Domestic practices such as wearing old clothes and leaving streets dirty and smelly were considered by the women a 'private language' to testify to the violence that had taken place but was inadequately addressed by the state. She argues that these micro-practices of mourning had special significance for these women, as the state was unable to provide an official public space where people's narratives and protests could be shared and perhaps addressed by the authorities. For Das, ethnographic research is a way to provide an 'occasion for forming one body, providing voice, and touching victims, so that their pain may be experienced in other bodies as well' (196). She argues that in cases where the testimonies of victims are unheard by the state and where legal and psychological discourses rob 'the victim of her voice and distance us from the immediacy of her experience', it is the task of researchers to provide an 'anthropology of pain' which views suffering as moral commentary and political performance (175). This ethnographic study takes up Das' advocacy and aims to give voice to sufferers frequently discussed but rarely heard both in Western debates about distant suffering and Filipino discussions of masa television audiences.

The empirical scholarship reviewed above has indeed provided rich accounts in their focus on different victims of exclusion in public life. It must be noted, however, that mention of media and their role in the lives of sufferers is glaring in its absence. For instance, although Das' study forcefully argues for the need of a public space in recognizing people's pain, the media are still left out of her discussion.

Aside from these studies, there are also more sociological approaches where the focus shifts from victims' lives and experiences to the social processes that lead to exclusion (Alcock 2006, 114). Whereas the first approach has been at times criticized for romantic and exaggerated readings of resistance in vulnerable peoples' everyday practices, the latter approach intends to criticize structural forms of domination and victimization.

The corpus of scholarship of the sociologist Skeggs (1997; 2004; Wood and Skeggs 2009) is a touchstone here. Her ethnography with white working-class women in Northwest England argued for the continuing significance of class as a point of analysis in sociology. She finds that in academic – and popular – discourse, there has been a movement to deemphasize class (Skeggs 1997, 6–7). However, she forcefully argues that the 'hidden injuries of class' (Sennett and Cobb 1973) still remain. For one, Skeggs found that working-class women's lives are circumscribed in an 'emotional politics of class fueled by insecurity, doubt, indignation and resentment [...] They were never able to feel comfortable with themselves, always convinced that others will find something about them wanting and undesirable' (Skeggs 1997, 162). Building on Bourdieu's (1986) framework of class as an accumulation of different forms of economic, social, cultural and symbolic capital that can be converted from one to the other, Skeggs traces how class is policed in different contexts, with working classes pathologized as having the 'wrong' kind of cultural capital and, subsequently, low moral value. Judgments of propriety and respectability in parenting, expression of emotions and display of the body, among others, are all classed, she argues, though usually these are dangerously constructed by society as individualized problems rather than social ones rooted in structural inequalities (Skeggs 2004, 60). Her work on reality television, discussed further in the next chapter, expounds on how its tenets of self-responsibility and self-management become mechanisms by which class inequality is reproduced by its imposition on working-class individuals to 'take up a particular form of agency, one to which they do not have access in the first place' (Skeggs 2004, 61; see also Wood and Skeggs 2009). Skeggs' idea that class is inscribed on the body resonates with observations from Das (1995), who also describes that distress, hurt and pain are always exhibited through the body and that these could become either a currency for recognition or a mark of shame.

Another key work in sociology that I draw on is Sayer's *The Moral Significance of Class* (2005). Sayer proposes a research agenda of understanding lay normativities with regard to class, that is, how people's moral evaluations of self and others are related (or not) to their class. He argues that the emotions of compassion, shame, envy and benevolence are not 'mere feelings or affects', but are expressive of moral evaluations of right and wrong. Sayer stresses that sociological analysis should be attentive to emotional responses and describe

the underlying normativities that people have in expressions of disgust or pity.
Further, Sayer emphasizes the moral aspects of the politics of recognition and
the politics of redistribution (Fraser 1997; Fraser and Honneth 2003). He ends
his book with a normative call:

> We need not only a politics of recognition but a rejuvenated egalitarian
> politics of distribution that confronts the injustice of class inequalities
> openly, one that does not treat class as if it were merely the outcome of
> competition amongst adults on a level playing field, but which recognises its
> profound effects on people from birth. In fact, redistribution in itself would
> be an advance in terms of the politics of recognition (Sayer 2005, 232).

Indeed, both bodies of literature that I have reviewed so far emphasize both
politics. Those that treat suffering as the problem of the witness are typically
concerned about the potentials and pitfalls of a politics of redistribution.
Those that deal with suffering as the problem of the sufferer typically describe
in ethnographic detail the struggles for recognition of sufferers themselves
whilst describing their desire for a fairer and more equitable politics of
redistribution. This book, as it examines people's lay moralities towards
mediated suffering, intends to be attentive to both sets of moral debates, as
it assumes that audiences, especially in non-Western contexts, are situated in
diverse social positions and may occupy either (and both!) positions of witness
and sufferer in the face of televised suffering. Following the perspective of the
anthropology of moralities, this study seeks to analyse the moral discourses of
right and wrong that are articulated in local contexts and how these discourses
are shaped by both social factors as well as the media themselves.

The problem of the sufferer in the Philippines

As mentioned in the introductory chapter, Philippine social science has
been historically described (and criticized) as being a 'sociology of coping
mechanisms' for its traditional emphasis on strategies of survival and resistance
in contexts of poverty, natural disaster, colonization, migration, illness, sexual
deviance and unequal or unjust systems of labour/exchange, among many
other contexts of suffering (David 2001a[1976], 42). David describes that
most of these studies employ the method of ethnography in order to expose
the creative workings of poor or marginalized communities.

An important recent work that falls under this rubric is Bankoff's (2003)
Cultures of Disaster. Mixing both historical analysis with anthropological and
geographical insights, he argues, 'natural hazards occur with such historical
frequency that the constant threat of these has been integrated into the

schema of daily life to form what can be called *cultures of disaster* [emphasis in original] (7). Bankoff challenges how natural disasters, hazard, vulnerability, risk and trauma have all been conceptualized from a Western perspective. Western thought would typically assume a 'myth of a secure and productive "ordinary life"' that is disrupted only during rare moments of disaster (159) and employ disaster management strategies using 'culturally defined technocratic notion[s]' with little sensitivity to cultural contexts (182). He cites the Philippines as an exemplar of a culture that has come to terms with 'living in the shadow of the volcano' (153) and excavates multiple examples of 'historical adaptations' and 'coping mechanisms' that Western studies and development organizations have failed to give credit to. For him, the 'simple *nipa* and palm hut' – a type of dwelling made out of branches and leaves from trees used in low-income or rural settlements – should not be vilified as 'primitive' but instead be perceived as adaptive, since these anticipate easy rebuilding after typhoons and earthquakes (163–4). He also argues against stereotypes that Filipinos are purely fatalistic or irrational in their decision-making, such as when a 1994 ferry tragedy took place due to passengers' conscious rejection of a Coast Guard warning to avoid sea travel during a typhoon. For Bankoff, Filipinos draw from folk, supernatural and religious resources to interpret and predict the work of nature and conceive of their fate as a product of the best efforts of divine and human intervention (168). He also narrates how bayanihan – systems of exchange and mutual assistance – are practised in communities. He argues that bayanihan is less an essentialist cultural trait than an adaptive behaviour that ensures future-oriented helping and gift-giving tied to traditional notions of debt and the nature of life as an endless cycle of prosperity and suffering (168–9). Bankoff's work is important for my study, for it describes how suffering might not be interpreted according to Western expectations and inflections of 'shock effect' (Chouliaraki 2010, 111–12; Cohen 2001, 203) but may have culturally informed judgments of disasters and the people affected by them.

Another ethnography that accounts for the variety of everyday practices and formalized rituals wherein people attempt to gain power is Cannell's (1999) study of 'those who have nothing' in a Bicolano city. From arranged marriages by which poor families extend their kinship networks for economic gain to healing rituals where the sick invoke protection from God and unseen spirits, Cannell describes how her poor informants constantly negotiate their low status and attempt to transform their life conditions. She states that everyday talk and local rituals are rife with 'idioms of pity, persuasion, and reluctance' that the poor use to strategically claim recognition from those with greater means (254). Cannell's work is relevant to my study in that it highlights acts of claiming pity, asserting victimhood, displaying reluctance and reminding the privileged of their moral obligations as practices worthy of critical reflection. How and whether such actions on the part of the poor become effective or not when situated in the media are interesting questions to consider, given how different television genres make visible performances and narratives of suffering.

Nevertheless it must be said that both Bankoff's and Cannell's studies can still be criticized by David's pronouncement that they overemphasize coping mechanisms with insufficient sociological critique of the causes of people's deprivation (David 2001a[1976], 42). David's call for critical analysis that takes into account issues of, for example, conflicting class interests or repressive global (or American) economic policies and trade practices (44–6) is rarely taken up even in recent research on the Filipino poor – with a few exceptions.

The anthropologist Pinches focuses on class conflicts between the poor and the middle and upper classes (Pinches 1987) and between the new rich and old rich Manila societies (Pinches 1999). Although not an anthropologist of moralities, he finds, as Bourdieu (1986) did in France, that different classes deploy different discourses of respectability to accord value for themselves: middle classes assert themselves as enterprising and creative whilst upper classes assert their heritage and their accumulated cultural capital to fashion themselves as superior to others (Pinches 1999). Official nationalist discourses of progress, in the meantime, are challenged by the poor, whose skepticism is borne of their experiences of exclusion by the rich who 'can't be bothered with us' (Pinches 1987, 101). Pinches' insistence on a continued criticism of class politics in Filipino society inspires my research on televised suffering in the Philippine context. How the upper class engages with or disengages from stories of suffering that they see on television as well as the visceral reality of suffering in the city would be of great interest in this study attendant to conflicting classed moralities and their consequences on social, political and economic inequalities in Philippine society.

Summary

This chapter has reviewed issues and concepts in three bodies of literature that are of significance to this book: the media ethics literature, the anthropology of moralities literature and the suffering literature (itself divided into debates about witnessing suffering and debates about the coping mechanisms of sufferers). My assumption is that investigating suffering in the Filipino context requires reflection not only on distant suffering, conceptualized as exceptional events that involve moral reflection for the witness of the disaster. Rather, a study on suffering here also requires exploration of its mundane and everyday experience by those who have nothing. It also requires sensitivity to local cultural understandings of suffering as well as an analysis of the relationships that sufferers have with their more privileged others. Class therefore is an important analytical tool in that it challenges us to explore the interactions between witnesses and sufferers in Philippine society and how the media might (or might not) function as a desirable space for cross-class understanding, compassion and charity.

The next chapter develops a more coherent framework that will attempt to link these three bodies of literature using the theory of mediation.

Chapter 2

THEORIZING MEDIATED SUFFERING: ETHICS OF MEDIA TEXTS, AUDIENCES AND ECOLOGIES

> The media do not simply add a new element to the story, they transform it.
>
> Sonia Livingstone, *On the Mediation of Everything*

As mentioned in the introductory chapter, to better understand the context of encountering suffering on television, it is important that we investigate audiences in their everyday lives. A focus on audiences can verify or challenge assumptions made in the traditionally Western-centric and highly normative literature on distant suffering and media witnessing. As discussed in chapter one, audience-centred research can explore tensions between normative theory and plural and culturally informed lay moralities. In particular, it is also able to capture the naturally-occurring responses of different people in relation to the poverty of television – both their specific interpretations of representations of poverty as well as their own lay moralities of good and bad media conduct in their very process of representation.

This book proposes however that we focus not solely on audiences. Rather, it is crucial to attend to the relationships that audiences have with media texts and media institutions at large. This book adopts a conceptualization of media as a process of mediation, one that accounts for the 'circulation of meaning' across different moments (Silverstone 1999, 13). Understanding media as a process requires extending the remit of analysis beyond the text or even the point of contact between text and audience and considers the dynamic nonlinear 'circuit of meaning' beyond the traditional model of producer/text/audience (Livingstone 1999). This requires me to draw on existing work in media studies about mediation (Chouliaraki, 2006; Couldry, 2000, 2008b, 2012; Livingstone 2009; Madianou 2005b; Silverstone 1999, 2005, 2007;

Thompson 1995) and argue for its significance as a theoretical guide for a critical analysis of media in late modern society.

Whilst the concept of mediation has been often used to describe how media logic has radically altered the conduct of politics (Couldry 2009; Silverstone 2005), its application to the study of the consequences of media to the experience of suffering potentially points to new ways of thinking about media ethics and audiences' responsibility to vulnerable others.

In addition, due to its attentiveness to the distinctiveness and interrelationship of moments of text/production/reception, mediation theory can challenge scholars to articulate more clearly their specific normative positions about the ethics of media texts, media production and audience reception. Current trends in media ethics, as articulated by the popular concept of media witnessing reviewed in this chapter, often fixate on the text and its ability to constitute moral audiences, with less to say about particular audiences' decisions to ignore or engage with televised suffering as well as the important ethical issues that arise in the process of media production. Indeed, with its methodological preference for ethnography, mediation can avoid the textual determinism as well as moral universalism of existing studies on televised suffering by foregrounding the diverse contexts in which suffering is transformed by processes of media production and reception.

This chapter develops a working approach for using mediation theory in specific relation to the media ethics debates. Reviewing a variety of theoretical, textual, experimental and ethnographic studies, I map out media ethics debates alongside key moments of the mediation process. I call these three categories textual ethics, audience ethics and ecological ethics. My broad argument here is that studies about distant suffering provide valuable material to ground the more abstract, normative debates about relating with the other in the context of globalization, broadcasting and media saturation (e.g., Peters 1999; Pinchevski 2005; Silverstone 2007). However, I discuss that these approaches currently lack holism in their critique by fixing their analyses on only one moment of the mediation process whilst speculating about their consequences to the other moments. In the section 'A Critique of Media Witnessing', I argue that although the concept of media witnessing attempts to account for the ethical questions in the production and reception of suffering, it commits the same oversight as previous studies in its similar inattention to the socio-historical conditions of the witness that it theorizes about. To address this gap, I outline the key tenets of mediation theory in the section 'The Mediation of Suffering' and suggest the kind of theoretical investigations and additional ethical questions it will pose about the representation and reception of distant suffering.

A Critique of Media Witnessing

One theory increasingly used in the media ethics literature to describe and critique the representation and reception of distant suffering is that of media witnessing. Initially used by Ellis (2000) and Peters (2001) to discuss epistemological questions about the representation of reality and the condition of spectatorship in late modern society, media witnessing has since been developed in the media ethics literature in judging the ethical practice of producers and audiences.

Through discursive critique of news reports, documentaries and fiction films, witnessing theory asks whether or not these individual texts successfully or unsuccessfully bear witness to events of trauma and suffering, such as the Holocaust or 9/11 (Frosh 2006; Frosh and Pinchevski 2009). Successful witnessing texts are measured by how they enable witnessing subjectivities through aesthetic, narrative and technological techniques that enable reflexivity, estrangement and an experience of loss on the part of their readers and viewers (Brand 2009).

At the same time, and sometimes rather confusingly, witnessing theory is also used by some scholars to discuss the experience of the viewers at home and processes of judgment that are provoked in the act of being eyewitnesses of televised events. Ellis' essay phenomenologically describes the condition of 'mundane witnessing' as a kind of 'default' experience of a paradigmatic (and de-historicized) television viewer. Whilst mundane witnessing involves a general 'awareness of events around us and of the people who make up our society and wider world', it 'does not require the detailed recall of news stories' nor any other kind of political or moral action (2009, 83). As with the majority of studies on media ethics, mundane witnessing assumes that the television viewer's ordinary life is one of safety and stability defined by geographic and social distance from suffering. For Ellis, this default position of a television viewer gives way to a more engaged and attentive audience experience only during events that may be 'traumatic in their implications for normal states of awareness or may bring up painful personal associations and deep fears' (2009, 85).

In addition, Peters' seminal essay titled 'Witnessing' foregrounds the problem of truth or authenticity as the central question that underlies the judgment of the viewer of mediated distant events. For Peters, audiences are involved in working through 'the veracity gap' that underlies their experience of being distant in space and time from events broadcast on television. Whereas doubt and distrust are present in the media witnessing of 'reports from distant personae' in contrast with interpersonal communication with '[those people] we know and trust' (2001, 717), he argues in some contexts such as those of *live* media events, doubt and distrust are reduced though the experience of '"being there" in time' (2001, 719).

Whilst the moral and epistemological questions raised by witnessing theory are often novel and compelling, the literature suffers from some ambiguity both in its confusing and inter-/ever-changing jargon and its unclear normative position about the moral obligations, if any, of the audience or the witness.

On the first point, the witnessing literature attempts to distinguish between passive and active forms of witnessing, such as the distinction we reviewed previously between 'mundane witnessing' and the more active, and the morally obliging, witnessing of traumatic events or events with personal resonances (Ellis 2009). However, such distinctions between passive versus active or amoral versus morally charged forms of witnessing often collapse or are even interchanged in the literature. Despite efforts to create typologies between 'eyewitness' and the person who 'bears witness' (Peters 2001, 709) or witnessing 'in', 'by' and 'through' the media (Frosh and Pinchevski 2009, 1), ultimately there is a lack of specificity and consensus in the literature as to whether 'witnessing' ultimately refers to (1) passive audience spectatorship or voyeurism, (2) active audience responses to media events of suffering, (3) textual strategies that lend authenticity to media representations of atrocity and therefore invite an active response from audiences or (4) all of the above. This confusion has led Tait (2011), in her own critique of the media witnessing literature, to propose a strict analytical distinction between 'witnessing' and 'bearing witness', where the latter should only refer to active, moral engagements with events of suffering. Tait's own empirical work, however, focuses on the practices of bearing witness by a particular journalist, with less to say about the diverse ways that audiences at home may (or may not) be able to bear witness.

Certainly, the question of how the audience may or should bear witness is another unresolved issue in the media witnessing literature, as there seems to be an overwhelming emphasis placed on the moral force of witnessing texts over their audiences. This is seen, for example, in Frosh's (2006, 274) writings, where 'bearing witness' is an 'act performed not by a witness but by a witnessing text'. In our previous review, it was evident in Peters' analysis that the textual feature of liveness in a television broadcast is that which resolves the veracity gap that audiences supposedly work through in their reception of television, assuming a rather linear relationship between a textual characteristic and an audience response. Most accounts of witnessing would have little to say about how differences in class, gender, age and ethnicity of audiences may affect the process of judging the authenticity of representations. And although Ashuri and Pinchevski's (2009, 146–7) framework of 'witnessing as a field' initially acknowledges the 'particular cultural boundaries, ideological settings and power relations' that shape audiences' judgments about distant suffering, their subsequent application of this framework in critiquing how documentaries bear witness to a West Bank conflict focuses completely on the textual elements of the documentaries while keeping silent about the audience. Indeed, long-standing arguments about the diversity of the audience and their interpretive skills (Ang 1996; Bird 1999; Press 1999) would challenge the presumptuous account of the activity of the witness in most of the literature.

As I elaborate in the following, mediation theory stands in contrast to this as it acknowledges the diverse positions of audiences who may orient themselves to media in different ways, even prior to the immediate encounter of the receiver with a media message. In its recognition of the distinct moment of reception in the mediation process, a mediational approach challenges media ethics scholars to articulate a clear position on the obligations, if any, they wish to place on audiences (or specific groups of audiences) in, for instance, seeking information, donating to or volunteering for victims on television or challenging stereotypical media representations.

The Mediation of Suffering

Sonia Livingstone (2009, 7) defines mediation as the ways 'the media mediate, entering into and shaping the mundane but ubiquitous relations among individuals and between individuals and society'. It emphasizes on the one hand the contemporary condition of media saturation, the fact that media are ever-present in modern life; on the other hand, and more significantly, mediation theory underscores media's transformational capacities toward social processes (Couldry 2008b, 380).

Methodologically, mediation theory pushes for a greater degree of holism in media research through its invitation to imagine relationships between production and reception. It hearkens back to Stuart Hall's (1999) 'circuit of culture', but requires researchers to simultaneously embrace and transcend the traditional model of producer/text/audience. This is why Livingstone (2009, 8) argues that mediation studies cannot simply use textual analyses or even the focus groups of traditional reception research; mediation theory invites an ethnographic methodology. Ethnography is viewed as the best methodological approach to follow the trail of media power and the circulation of meaning across different moments of the mediation process. Studying the mediation of distant suffering then begins with the everyday lives of a group or groups of audiences in a particular locale, identifying the programs and texts of suffering that they encounter, analysing the textual characteristics of these texts and recognizing how particular audience responses and interpretations are affected by the text and other social factors that prove salient during ethnographic observation.

In other words, mediation requires the analysis of the actual consequences of media and communications technologies on the reality of distant suffering. This necessarily engages with questions of representation raised in the textual ethics and media witnessing literatures to be discussed later in this chapter, including whether sufferers are portrayed with humanity and agency (Chouliaraki 2006; Orgad 2008) and whether the account of the event is truthful and believable (Peters 2001). This analysis will, however, treat these questions not in isolation (i.e., through internalist critique of texts or technologies), but as immediately interrelated with their reception by audiences – what I refer to as audience ethics. Therefore, ethnographic interviews and participant observation with audiences are indispensable to a mediational approach, as it is through audience analysis that linkages are made between the visual and rhetorical techniques of representing suffering and audience responses of turning away, expressing discourses of compassion, or donating and volunteering (insights from audience ethics). Mediation theory must necessarily be sensitive to qualify how particular techniques of representing suffering may evoke different responses to audiences according to categories of class, race, age, ethnicity, or individual experience, recognizing that 'the social is in turn a mediator' (Silverstone 2005, 189).

Additionally, it will invite reflection on the ethics of the media process, particularly in the interactions between media producers and the sufferers that they represent, which I call ecological ethics. This follows from mediation theory's expansive conceptualization of media power as including people's 'direct experiences with media' (Couldry 2000; Madianou 2005). In so doing, mediation may challenge the orthodox literature in its sensitivity toward the perspectives of those represented by recording how sufferers regard their own representations as well as their personal interactions with media producers. Following these principles, it is indeed possible to dialogue back to the normative (and sometimes speculative) positions of researchers about agency and compassion fatigue by qualifying how these concepts hold up in local contexts.

The use of mediation rather than its related (and often interchangeable) term mediatization is intentional here. Although there are clear overlaps of meaning with both these terms that seek to describe media's cumulative effect on social organization (Lundby 2009), the emphasis on conversation and mutual shaping between media and audiences that has been traditionally inscribed in the concept of mediation offers a more useful and clearer direction for empirical research on distant suffering. Whereas mediatization directs researchers to seek out the media logic of a text or medium and examine how it recalibrates the operations of a particular social process (Altheide and Snow 1979), a direction that has certainly enhanced our understanding of how political campaigns have been transformed by television's logic of entertainment (Meyer 2002), it is perhaps less useful in the context of studying distant suffering and media. In this context, identifying a single media logic that underpin the multiplicity of genres and narratives that represent different events of suffering and the different ways in which producers and journalists interact with sufferers is not only analytically boastful but perhaps even unnecessary if our interest is really to nuance and specify the generalities previously laid out by works on media witnessing and compassion fatigue.

In the next sections, I situate current media ethics debates within three different frames of analysis consistent with a mediational approach. My discussion groups theoretical, textual, experimental and audience studies along a three-fold typology that illustrates ethical concerns about mediated suffering in relation to three different moments of the mediation process. I call these categories textual ethics, audience ethics and ecological ethics. In the section 'A Critique of Media Witnessing', I argue that although the recent development of media witnessing theory attempts to account for the ethical questions in the production and reception of suffering, it commits the same oversight as previous studies in its similar inattention to the socio-historical conditions of the witness that it theorizes about. Finally, I highlight the specific contributions of a mediational approach to the study of televised suffering.

Textual Ethics

Studies classified under textual ethics are concerned with the textual politics of the representation of suffering, particularly the analysis of the moral claims embedded within individual texts or groups of texts. The methodological tools used here are content analysis, discourse analysis, visual analysis and general impressionistic analysis of texts. A common assumption shared by scholars here is that media texts hold symbolic power in the ways in which they make visible the suffering of others and thus offer claims for compassion from audiences. As Chouliaraki (2006, 113) notes, 'The point of departure in reversing compassion fatigue is to actually tap into the texts of mediation and work on their pedagogical potential for evoking and distributing pity.'

Perhaps the significant contribution of works in textual ethics to the broader media ethics literature is their skillful distilling of abstract philosophical principles down to the critique of individual, instantiated and intentional objects of media production. For instance, while the appropriation of the moral philosophy of Emmanuel Levinas in media studies (discussed in chapter one) began with philosophical essays that discuss the ethical challenges of interacting with the other in the space of technologically mediated communication (Peters 1999; Pinchevski 2005; Silverstone 2007), the analytics by which one could identify the particular processes of othering in media narratives and media spaces is developed by text-centred studies. Indeed, in a recent special issue in the *International Journal of Cultural Studies*, Silverstone's philosophical prescription about the media's infinite responsibility to the other, particularly through his normative concept of 'proper distance', is applied and developed through textual critique of the news (Orgad 2011), humanitarian advertising (Chouliaraki 2010) and accounts of torture (Sturken 2011).

As discussed in chapter one, a key concern for many scholars here is whether or not sufferers should be represented as possessing agency, that is, whether they should be shown to have the capacity for self-determination and independent action. For some scholars (Moeller 2002; Tester 2001), media narratives that depict sufferers as possessing agency are assumed to enable (Western) audiences to effectively relate to or identify with the situation of distant sufferers. They further argue that depictions of sufferers without agency are assumed not only to disable identification but in effect also to reduce the humanity and dignity of the sufferers.

Chouliaraki's *The Spectatorship of Suffering* (2006) maps out a typology of news narratives of suffering based on the different ways that agency is conferred on sufferers by visual and rhetorical elements of news texts. The least morally desirable news narrative is what she calls 'adventure news', which depicts sufferers as having little or no agency by using 'dots-on-the-map' imagery and impersonal references to sufferers as aggregates of victims, thus containing no moral claim for audiences to care for sufferers. She contrasts this with the more morally superior techniques used by 'ecstatic news' and 'emergency news', which offer a more complex variety of positions for spectators to feel and act for distant sufferers (Chouliaraki 2006, 137–46). Visual techniques of using both long shots and close-ups and rhetorical strategies of giving a name to the sufferer are assumed to confer agency and humanize the sufferer.

While Chouliaraki uses discourse analysis in her book, other scholars rely on general impressionistic analyses of texts. For instance, Moeller observes and criticizes the 'repertoire of stereotypes' used in the news. Among these stereotypes are the portrayal of rescuers as heroes (1999, 43, 104) and the portrayal of sufferers as the 'starving innocent child' (2002, 53). She argues that representations of suffering using and reusing such conventions simplify the conditions of sufferers, and she speculates that this creates 'compassion fatigue' among audiences.

However, Orgad (2008) proposes a counterargument to the view that it is better and more ethical to depict sufferers as active agents. For her, depicting too much agency in sufferers may be detrimental to their cause, as viewers might see sufferers as in fact capable and independent and thus in no need of attention or aid. In her comparative content analysis of newspaper coverage of the October 2005 South Asia earthquake and the 7/7 London bombings in 2005, Orgad (2008, 18) found that the pain and suffering of the sufferers in the South Asia earthquake were considered to be 'more tolerable' and 'having feasible solutions' in comparison to the London bombings. Orgad observed that the word used to describe the sufferers in the South Asia Earthquake was *survivors*, while the term for London bombing sufferers was *victims*, even though the number of actual casualties of the South Asia Earthquake greatly exceeded that of the London terror attack. She argues, challenging Tester and others, that portraying sufferers as victims and portraying suffering at its worst may in fact convey 'a message of agony that requires political action' (2008, 21).

Whilst scholars in both camps are divided on the issue of whether sufferers should be depicted with more or less agency, many share the normative judgment that the media should attempt to trigger the compassion of audiences for 'distant' sufferers and not just 'our own'. One normative principle useful to recount here is Silverstone's (2007) notion of proper distance, mentioned earlier in the chapter. Proper distance is a guide for action in an intensely mediated world not only for media producers but also for media audiences. Proper distance is lost when the other is pushed 'to a point beyond strangeness, beyond humanity' or drawn 'so close as to become indistinguishable from ourselves' (2007, 47). Inspired by Arendt's notion of the in-between, previously discussed in chapter one, proper distance is a way of relating with the other neither as 'too close' nor too far'. This concept is then applied to critique media representations of the other as either too close (celebrities, the exotic in advertising) or too far (terrorists, Muslims) (2007, 48).

In addition, there are scholars who argue for the different affordances of not just discrete texts but entire genres when it comes to affecting audiences' moral responses to suffering. For example, Ignatieff (1998) is skeptical that the news genre can be an effective trigger for audiences' compassion. He claims that the short ninety-second slices of news 'dishonours' the world's horrors, whilst the best documentaries can achieve 'the prerequisite of moral vision itself', given how their length and structure enable contextualization of complex events (1998, 30–32). Recently, Chouliaraki (2012) and Seu and Orgad (2010) have also explored representations of suffering in humanitarian advertisements.

Genre is indeed an important concept to consider in studying the mediation of suffering. As I discovered during my fieldwork for this book, most of my respondents discussed the reality television genre when asked directly about where they most frequently encountered suffering on television. Genre, defined as 'systems of orientation, expectations and conventions that circulate between industry, text and audience' (Neale 1987, 19) is a concept that emphasizes the creative process of meaning-making that links the production and reception of texts based on shared semiotic and cultural expectations. In comparing audience responses to specific texts in news and factual entertainment genres, this book explores how genre differences enable different moral discourses to be expressed based on audiences' evaluations and expectations of genre conventions.

In spite of the many contributions of the textual ethics approaches to advancing discussions of ethics beyond the dulled yardsticks of 'objectivity' and 'impartiality' in studying the representation of suffering, these studies suffer from key limitations. First, textual ethics approaches can be criticized as suffering from a determinism that overstates the consequences of media texts on audiences. For example, Moeller's (1999, 2) thesis about compassion fatigue claims that audiences' disinterest toward distant suffering 'is an *unavoidable* consequence of the way the news is now covered' (emphasis added). Tester also assumes that the news genre 'allows the audience no time to spend with the suffering and misery of others, making it instead a fleeting concern' (2010, 50). However, it must be said that the hyperbolic determinism of Moeller and Tester is not shared by other scholars who acknowledge texts as performative of already-existing values and presenting proposals for ethical action and contemplation (Chouliaraki 2008).

Second, some textual ethics studies approach their critique of texts with an a priori assumption that the agency of sufferers is what shapes and influences audience responses. Between the disagreements whether it is the presence or the absence of agency that prompts 'political action' on the part of the audience (Orgad 2008, 21), what is set aside are factors beyond agency that may turn out to be more significant in shaping audience response. The site of suffering may in fact be the common ground between Chouliaraki's and Orgad's analyses, as they similarly identified greater public attention accorded to Western atrocities, 9/11 in the United States (Chouliaraki) and 7/7 in the United Kingdom (Orgad), in spite of varying degrees of agency they identified between the two. In addition, we can also speculate that the cause of suffering could have affected audience responses more than the presence or absence of agency, given that their examples identified terror attacks rather than natural disasters or mundane poverty as receiving greater attention.

Indeed, textual ethics studies have more productive contributions in their distilling of ethical principles from their judgment of representations of suffering rather than in their speculative accounts of how good or bad texts are linked with active or passive audience responses. In the next section, we turn now to the contributions and limitations of audience-centred studies to the ethical debates about the reception of distant suffering.

Audience Ethics

Audience ethics in relation to distant suffering includes studies of both consumption and reception of audiences. Whereas the former focuses on access, preference and interest in particular platforms or programs, the latter foregrounds the emotional responses and interpretations that people have about specific narratives of suffering.

It is crucial to discuss first the normative position of many researchers that audiences should know about the suffering of others. Ignorance of others' suffering or, in more general terms, ignorance of public issues is considered less desirable than being aware and attentive. This expectation on audiences, in normative terms referred to as 'publics' (Livingstone 2005), explains why audiences' consumption of news is a significant point of study for many scholars. News watching is assumed to be a civic duty for audiences, because it is considered a crucial practice in the operations of democracy.

In the discussion of distant suffering, scholars in different fields of media studies, sociology and social psychology express a moral evaluation that audiences should act as 'moral spectators' (Boltanski 1999), rather than disinterested 'metaphorical bystanders' (Cohen 2001, 15). While it is recognized that there is 'no decent way to sort through the multiple claims on our time or philanthropy' in the face of the world's atrocities (Midgeley 1998, 45–6), scholars nevertheless positively value audience activities of seeking information (Kinnick et al. 1996), empathizing and analysing (Donnar 2009), donating money (Tester 2001) or even the mere act of viewing rather than turning away (Cohen 2001; Seu 2003). Luc Boltanski (1999, 18–19) rescues the value of speech and protest as legitimate actions toward media narratives of suffering, contrary to perceptions that talk 'costs nothing' or has no consequence. He suggests a modest, 'minimalist' ethics where 'effective speech' is viewed as a valuable moral action for spectators of distant suffering.

Just as in textual ethics, compassion fatigue is a recurring term across this literature, as different scholars are concerned about patterns of society-wide desensitization and indifference to social suffering as a function of mediation. However, compassion fatigue is operationalized and measured in different ways. There are those who empirically study patterns of avoidance toward televised suffering (Kinnick et al. 1996); some research rhetorical responses of apathy or pity toward specific texts of suffering (Höijer 2004); others theorize about both (Cohen 2001; Seu 2003).

Kinnick and his colleagues' study claims to be the first empirical investigation of compassion fatigue as it relates to media coverage of social problems. They argue that compassion fatigue can be measured in people's selective avoidance of particular issues in the news. They say that an issue that is emotionally distressing for an individual 'is more likely to be associated with avoidance behaviors, ostensibly as a form of self protection' (1996, 700–1), while issues of interest are associated with 'information-seeking' (1996, 698). By using telephone survey methodology, however, their study was unable to explain the reasons why certain issues prompt more or less avoidance and was likewise unable to tease out the specific textual strategies of the specific media reports that trigger avoidance.

Seu's focus-group-based social psychological study delves deeper into this issue of avoidance. First, she argues that compassion fatigue is not a result of information overload or normalization, but is in fact an *'active* "looking away"' [emphasis in original] (2003, 190). Her interviews uncover that participants routinely used clichéd psychological terms such as 'desensitization' when talking about why they turn away from humanitarian advertisements. Crucially, Seu (2003, 190–2) argues that desensitization is not an explanation but a moral justification; popular psychological discourse becomes a resource that people draw upon to distance themselves from their responsibility to others' suffering. While Seu's critical approach is useful in the ethical critique of audience responses, particularly in its clear normative position that compassion fatigue is an individual moral choice rather than a top-down social or historical process, its limitation lies in its inability to link the individual moments of turning away with the specific visual or rhetorical prompts that might trigger these undesirable actions.

One of the most-cited audience studies on televised suffering was Höijer's work in Norway and Sweden. Using both surveys and focus groups, she enumerated different 'discourses of compassion' that her respondents expressed toward distant suffering (2004, 522–3). Höijer's sociological contribution is her observation that these expressions appear to be gendered: females are more likely to express compassion while men 'shield and defend themselves by looking at the pictures without showing any outer signs of emotion' (2004, 527). Nevertheless, she fails to elaborate on how different discourses of compassion might be expressed toward particular kinds of news clips or genres, or toward representations of suffering with various degrees of agency or causes of suffering or how people make judgments about the media's actual role in representing suffering. Finally, although she highlights gender differences in her sample, she neglects the salience of other categories such as class, age, religion and even ethnicity, in spite of her sample coming from different countries in shaping the experience of witnessing distant suffering.

Seu's insights resonate with the sociological work on denial and emotion management carried out by Norgaard (2006), who studied social movement non-participation in Norway. Inspired by Hochschild (1983), Norgaard's ethnography found that people held conflicted feelings about their perceptions and practices of caring about the environment. People felt fearful and guilty when thinking about environmental issues, but instead of engaging in social action, her respondents were seen to manage these negative emotions.

They controlled their exposure to information in order to deflect negative feelings of fear and guilt (2006, 389–91).

Findings in audience ethics on 'switching off', turning away and denial are relevant to my study, as the Philippine literature I reviewed in the last chapter hinted at classed strategies of denial, such as 'flying over' (Tadiar 2004), 'cutting the network' (Strathern 1996; McKay 2009) and the elite's distancing from the poor to reduce obligations (Kerkvliet 1990). The critical approaches of Seu (2003) and Norgaard (2006) enable me to analyse how responses to suffering may in fact be coated as justifications in the context of a moral dilemma interview. But unlike Seu, I aim to provide a more sociological than individualistic analysis of how these moral discourses may be rooted in particular classed, gendered, religious etc. positions. And unlike Seu and Norgaard, I hope to theorize how the media themselves are implicated in people's consumption, or avoidance, of televised suffering. Indeed, much of the work on compassion fatigue reviewed here assumes that compassion fatigue is a consequence of audiences' declining concern for others' suffering, rather than an active response to media practices of representing and interacting with sufferers. It is crucial for us to explore in this book how audiences' discourses of compassion toward sufferers on television might be dependent not only on an evaluation of sufferers themselves but on what I call 'lay media moralities' about the very process of mediation, which pertains to audiences' judgments on whether media are exploitative or helpful, manipulative or sincere, in their interactions with sufferers.

A reception research approach is seen in Kyriakidou's focus groups with Greek audiences of televised suffering, where she found that older and lower-income people were more able to expression discourses of compassion for distant sufferers. These groups tended to have more first-hand experiences with suffering than younger and more affluent groups; the common experience of suffering enabled expressions of empathy, for example, '[we] identify more with the poor [...] because, unfortunately, this is who we are as well' (Kyriakidou 2005, 162). Meanwhile, middle-class audiences could not or were even 'unwilling to emphathise' (2005, 165), though Kyriakidou did not explain why this is so. Nevertheless, these classed patterns in audience responses to suffering will indeed be interesting to explore in the context of a mediational project on televised suffering, as it foregrounds the interactions of social factors and media narrative in the moment of reception.

Using reception analysis themselves, Skeggs et al. (2009) expand too on the insight that audiences distance themselves from texts by further arguing that this rhetorical ability is a resource possessed by middle-class and not by working-class audience. Following Bourdieu (186), the researchers argue that 'modes of articulation' are generated through available classed capitals. Middle-class audiences did not want to be attached to 'that which is a cultural display of working-class (low) taste. They need to show not only cultural detachment, but also cultural superiority to the bad object' (Skeggs et al. 2009, 11). Although Skeggs et al. focus only on the genre of reality TV, we can extract interesting questions about the mediation of suffering from their work. In their insight that audiences have classed moral judgments about media genres and practices, we can explore how audiences also express lay moralities about media practice, evaluative statements of good and bad in relation to media conventions of representing suffering as well as their general expectations toward journalists, producers and audiences like them. Indeed, one issue that is unresolved in the distant suffering literature is whether audience decisions to seek information or look away are primarily issue-driven (Kinnick et al. 1996), or prompted by judgments about how media represent these issues (Cohen 2001; Moeller 1999) or by a general distrust toward media and other social institutions. This idea that audience responses to suffering may be informed by knowledge and judgment about the media themselves links with the next section on ecological ethics, a third category of media ethics that we should consider alongside textual ethics and audience ethics.

Ecological Ethics

As mentioned, the ethnographic perspective of mediation studies widens the remit of analysis beyond the text and beyond the encounter between text and audience; it also includes people's direct experiences with the media frame. Couldry (2000) and Madianou (2005b) have previously attested to how people's direct contacts with journalists and media institutions inform their interpretation of individual television texts. And in the wider media ethics literature, writings about the 'media environment' (Silverstone 2007), 'media ecology' (Born 2008; Born and Prosser 2001), and the 'mediated center' (Couldry 2003) have raised normative questions about the media at the infrastructural, institutional and organizational scales, which I would categorize here as ecological ethics.

A mediational approach therefore can lead us to connect the ethical debates about distant suffering in textual ethics and audience ethics to the ecological ethics of how media institutions should create an ethical and democratic space that upholds equality of voice (Couldry 2003; Cottle 2006), hospitality for vulnerable others (Silverstone 2007) and the just treatment of employees and participants (Mayer 2011; Ouellette and Hay 2008; White, 2006; Wood and Skeggs 2009). Although these writings variably discuss ethical concerns across different scales, from system-level critiques of media ownership to organization-level critiques of media production processes, the common focus here is on how control over media representation may be democratized and diversified rather than concentrated to economic and cultural elites in society. Granting mediated visibility to the invisible and voice to the voiceless are considered social goods (Cottle 2006, 175–82), just as the abuse and humiliation of participants and employees in the creative industries are critiqued. In the context of distant suffering, then, I identify that the more expansive discussion of media exploitation in ecological ethics can inform the text-centred debates about how media should (or should not) grant agency to sufferers. Particularly, a mediational approach can link the critique of the intentional object of representation (the 'moment' of the text) with the experiences of all those involved in the process of representation (the 'moment' of production), including the experiences of the sufferers recruited and interviewed by television producers.

Studies that shed light on the recruitment of participants in reality television (White 2006; Wood and Skeggs 2009), news (Madianou 2011) and fictional drama (Mayer 2011) raise ethical questions so far absent in the distant suffering literature. For instance, the commonplace judgment that to personalize and give name to the sufferer is ethical insofar as it humanizes her or his condition (Chouliaraki 2006) is complicated by a critique on whether the sufferers were given just (economic) compensation (White 2006), had control over their own speech and actions while on television (Wood and Skeggs 2009) or admitted to experiencing shame by being placed under the media spotlight (Madianou 2011). Certainly, judgments about the ethical value of representation from only a textual ethics approach might be contradicted by empirical findings that may indicate that those represented contest their representation and the process by which it took place. A more complex scenario is posed in Mayer's (2011) research on the filming of the HBO series *Treme* in a Louisiana town after Hurricane Katrina. Her study highlights how members of this traumatized community engaged in a form of self-exploitation by working as extras for long hours, but in fact articulated the positive symbolic rewards of appearing on television and commemorating their real experiences of loss. Approaching this case study using a mediational approach to distant suffering can provoke reflection about how to reconcile (if at all) the contrasting moral judgments that may arise from the different moments of text/production/reception. Does the moral judgment of the researcher supersede the lay moral judgments of audiences? Does a good, ethical text of suffering justify the bad, unethical process by which it is produced? How can we think about the agency of sufferers as involving not only textual characteristics but also the sentiments of 'those represented, not to mention the lay moralities of agency and respectability held by different audiences? These are difficult questions so far absent in the distant suffering literature and wider media ethics debates that are explored in this book.

Philippine Media Landscape

Having reviewed the media ethics debates about televised suffering and ethical questions across moments of text, production and reception, it is now important to gain insight into the Philippine media landscape.

The motivation to focus on televised suffering is partly explained by the dominance of television in the local media landscape and its significance in public life. Ninety-four per cent of the population has access to television, with radio, the second most pervasive medium, accessed by 60 per cent (AGB Nielsen 2010). A lifestyle habits survey by McCann-Erickson (2009) also indicates that watching television is the preferred leisure activity of Filipinos of various backgrounds, garnering an 82 per cent response rate. Listening to radio (52 per cent), chatting face-to-face (28 per cent) and Internet surfing (27 per cent) ranked markedly lower. Even with the growing availability of mobile phones and personal computers over a decade – with respective penetration rates of 100.3 per cent (Business Mirror 2013) and 36 per cent (Albert 2013) – television viewing levels have remained constant, with an unchanging average viewing time of three hours a day from 2001 to 2008 (McCann Erickson 2009), though other measures peg viewing hours at as high as seven and a half hours a day (AGB Nielsen 2010; Burgos 2010).

Two privately-owned television channels dominate Philippine television today: ABS-CBN and GMA Network. Though there exist state-owned channels, viewing levels for these channels are low, given that public channels have lower production and marketing budgets than private channels.

Prior to the Marcos dictatorship (1972–1986), a period when privately owned television channels, radio stations and newspapers were censored or sequestered for government propaganda, the Philippine media landscape had been renowned among its Asian neighbours for being one of the 'most democratic' and having 'one of the freest presses' (Coronel 2001; Rosenberg 1974). The Philippine press has been noted for its storied role in the nation's history, as it was through newspapers and periodicals penned by revolutionary leaders that the 'imagined community' of a Filipino nation was first born (Anderson 1983, 26–9). However, the martial law period under Marcos was as a period of trauma (Claudio 2010) to 'Asia's freest press', as censorship and crony journalism dominated during this time. The post-martial law Constitution, used to this day, was developed to restore and ensure media freedom, as the media are understood here as the Fourth Estate (Smith 2000). The Constitution has no specific laws to regulate the press, and the libel and

Table 1: Q1 2009 Audience Shares by Channel

Channel	Audience Share	Ownership
GMA Network	43%	Private
ABS-CBN	31%	Private
All Cable	9% (divided across 70+ cable channels)	Private
TV 5	8%	Private
RPN	4%	Public (privatization in process)
QTV Channel 11	3%	Private (GMA affiliate)
Studio 23	1%	Private (ABS-CBN affiliate)
NBN	1%	Public

Source: AGB Nielsen Q1 2009 Mega Manila Households Ratings Survey

sedition laws that would affect its performance were considered to be fairly liberal (CMFR 2007). Television networks can even evade sanction from self-regulatory bodies by freely withdrawing their membership (Ong 2010).

ABS-CBN and GMA Network today are as media conglomerates that extend their dominance across radio, print, cinema, the World Wide Web and even mobile services.[1] Like competing print titles and radio stations, they are owned by 'old rich' families with longstanding histories in local politics and big business (McCoy 2009).

ABS-CBN and GMA are together known as the network giants, embroiled in the most brutal ratings wars (Rimban 1999). The war here is a battle for the viewership of the so-called 'DE' low-income socio-economic class that comprises around 84 per cent of the population (AGB Nielsen 2010). Attracting the low-income class is actually deemed most lucrative for advertiser revenue,[2] as advertising spends are dominated by mass-oriented haircare, telecommunications and fast-food brands (McCann Erickson 2009). Though the upper class has understandably higher spending power per capita, the bulk of Philippine business is actually driven by lower-class consumption, given their sizeable proportion in the population. Marketers and academics refer to the Philippines being a 'sachet economy' (Arceo 2004; Tolentino 2011) – 'sachet' originally referring to the small sachet packages of personal care products such as shampoo, soap and toothpaste which are sold individually but now also indicating the purchase of small domination prepaid (or pay-as-you-go) mobile phone credit used for texting and calling.[3] The sachet economy business model of top advertisers (Unilever, Procter & Gamble and Smart Telecommunications) has been adopted by the broadcasting business, with publicly available terrestrial channels oriented to the lower-class demographic, and cable channels with mostly foreign programming[4] oriented to the upper and upper-middle classes.

Curiously, business concerns for profit are supposedly a recent phenomenon among television companies. In a 1996 interview, Bayantel, Eugenio Lopez III, the oligarchic owner of ABS-CBN, as well as the utilities firm Meralco and telecommunications firm, explains,

> You'd consider that [paying attention to ratings] pretty standard in the business, but the truth is, in the last 10 years, nobody has paid attention to the ratings except us. It is now only that [GMA Network] has woken up and begun to pay attention to ratings [...] Under martial law, all of these stations were dominated by political concerns, not by economic or business concerns [...] [Now] we've also paid attention to local[5] programming. We made a strategic decision earlier in the game that we were going to be a mass-oriented TV station (Romualdez 1999, 55).

As recently as the mid-1990s, television programming was not wholly directed to the masses: evening newscasts were delivered in English, American soap operas (aired in their original language) occupied important timeslots, and public affairs talk shows proliferated in response to post-martial law demand for political debate and 'revelation' (David 2001c[2000]).

Today both ABS-CBN and GMA Network are in heated competition for the hearts and minds of those they call the *masa* (mass) audience. Critics have deplored various trends in *masa*-oriented programming – from the 'soap epidemic' (David 2001c[2000]) to 'pied piper' game shows (Coronel 2006a; Doyo 2006) to the 'pornography of poverty' (Devilles 2008) – that entertains rather than educates or informs poor viewers. In contrast to the tendency of middle-class-oriented Indian television to be in denial of the widespread poverty in its country (Mankekar 1999; Sainath 2009), Philippine television constantly portrays scenarios and situations related to poverty. Soap operas typically have a young poor female character as their main protagonist, reality shows have poor elderly housewives as eager participants competing for cash prizes, and news broadcasts cover both momentous events of natural disaster and the everyday problems of slum communities.

Additionally, ABS-CBN and GMA Network have their own charity organizations in formalized synergy with their news departments. News correspondents arrive at sites for coverage of landslides armed not only with their broadcast equipment but with relief goods in the form of canned food, secondhand clothes and medicine from their charity organizations. ABS-CBN's *Bantay Bata* (Child watch) is integrated into the ABS-CBN News and Current Affairs Department. GMA Network's Kapuso Foundation is meanwhile headed by the most senior news personality of the network, Mel Tiangco. The synergy between the network's news teams and their charity organizations extends to the actual content of their primetime newscasts, as human interest segments feature donations by media charities to impoverished communities.

A pessimistic reading of this practice is that such strategies constitute ways by which patron-client ties are forged between television networks and their poor audiences. By providing direct assistance and solutions to their viewers' own problems, they cultivate personal relationships with their viewers in the hopes that recipients of charity may become loyal viewers of their channels. In the Philippines, after all, the majority of television audiences are not viewers of select programmes but of entire channels (GMA Research 2004), identifying themselves according to network taglines *Kapamilya* ('Of One Family', ABS-CBN's tagline), *Kapuso* ('Of One Heart', GMA Network's tagline) or *Kapatid* ('Sibling', TV5's tagline).

A more optimistic reading of this practice, certainly one preferred by the television networks, is that media fill the gap left by corrupt and bureaucratic public institutions – not to mention the absence of welfare state provisions. As GMA news anchor and Kapuso Foundation head Mel Tiangco explains in an April 2009 interview for this study to which this book is based,

> The reason why Kapuso Foundation is tied to the news is because, from our news desk, we learn days beforehand of the possibility of, say, a typhoon, and therefore we are able to prepare and help the victims. Why, who arrives on the site first anyway? It's always us [the TV networks]. And it's just irresponsible if we arrive and do nothing but just cover, when people are actually dying, or buried in [rubble] [...] But when Red Cross and the government and the other charities arrive, then we [Kapuso Foundation] would pull out. But Kapuso Foundation is always there as the first to respond.

This institutionalized synergy of charity and news is one characteristic peculiar to Philippine television, and absent in my review of other media systems. Locally this issue is one that has surprisingly received little critical reflection. Print journalists, usually quick to criticize their counterparts in television with 'dumbing down' accusations and complaints of sensationalizing suffering in entertainment genres (Contreras 2011; Stuart Santiago 2011), are generally accepting of media charities helping the poor. As we saw in chapter one, De Quiros (2009a), 'forgives' ABS-CBN and GMA for 'advertising their wares' whilst rescuing victims and distributing relief goods during Typhoon Ondoy.

Figure 1: A fisherman on Bantayan Island stands beside his new fishing boat donated by GMA Network in the aftermath of Typhoon Haiyan.

Photo by the author.

There has been little academic work done on television audiences in the Philippine context and even less on class differences in television consumption or reception. Cultural critics and opinion columnists note, however, that elites generally deride *masa*-oriented programming and celebrities as cheap and low-status, though they sometimes couch such judgments in less direct and offensive terms (Santiago 2010). Cordova (2011) claims that displays of poverty on television are practices that 'the Filipino elite always reviles'. He observes that the elite mentality is: 'We're not supposed to wash our dirty linen in public. We're not supposed to show our wretchedness.'

Flores (2001), however, has done rare ethnographic work in his study of (low-income) fan audiences of the celebrity Nora Aunor. Flores attests to the intimate relationship that female fans have with Aunor and her films. He coins the term sufferance – the patient endurance of suffering – to describe how Aunor's fans engage with Aunor's movies and fan events. Aunor's film persona is typically of the *atsay* (maid) who endures hardship and attempts to overcome poverty and tragedy. According to Flores, watching and talking about Aunor's movies with fellow fans enable them to express hope-oriented discourses about their own future conditions.

Conclusion

This chapter has provided a working model of mediation for a study on the ethics of televised suffering. Though primarily an audience-centred study, the media ethics debate on suffering alongside audience ethics (consumption and reception of audiences) and ecological ethics (the infrastructural, institutional and organizational scales of media), discussed in this chapter will be used to inform the analysis of audiences' responses. Following a mediational approach, which uses ethnography to examine the circulation of meaning, the analysis of domestic contexts of consumption and reception of television enables a better understanding of the tenability of normative categories in the context of everyday life. It also allows for an exploration of possibilities where culturally specific lay moralities may challenge universalist assumptions and normative claims. The following chapter takes us through the actual process whereby these lay moralities were encountered and the active role that I had in eliciting these from my respondents.

Chapter 3

AUDIENCE ETHICS: MEDIATING SUFFERING IN EVERYDAY LIFE

[What] has brought about the chaos and decay which has become the monument of the city [...] is the crisis of conflicting desires and social practices, caused by the intensification of the constitutive contradictions of capital.

Neferti Tadiar, *Fantasy-Production*

Drawing primarily from life story interviews and participant observation with upper-, middle- and lower-class people in Manila, this chapter analyses the moment of consumption in the mediation process. I specifically explore people's affinities towards local television and reflect on their significance to specific audience ethics debates about the significance of mediated engagement with the public world (discussed in chapter two). Whether people tune in or 'switch off' in relation to other people's suffering on television is regarded in the literature as a moral issue that audiences negotiate in their everyday lives (Cohen 2001; Seu 2003). And whether they seek out information, donate to charity or speak up against the suffering of others – 'even if this speech is initially no more than an internal whisper' (Boltanski 1999, 20) – are actions that scholars expect of an audience 'morally culpable' to 'take that responsibility' insofar as media enable connection (and disconnection) with vulnerable others (Silverstone 2007, 128).

   This chapter uncovers that these philosophical norms play out in a contested – and class-conflicted – milieu in the specific ethnographic context of the Philippines. It exemplifies the class-informed moral judgements about the 'poverty of television', where upper-class respondents' avoidance of over-representation of suffering and lower-class respondents' affective consumption of media content that reflects the hardship of their lives are ultimately expressive of classed moralities about media. I demonstrate in this chapter that media consumption here is reflective of, even an amplification of, class inequalities as elites turn to international and particularistic media with minimal consumption of 'depressing' Philippine media. Lower-class audiences meanwhile turn to local media not just for entertainment or information but also for the material rewards – and social services such as health care and legal advice – that television networks offer along with their televised content. A class-divided metropolis, with parallel and occasionally intersecting 'zones of safety' and 'zones of danger' (Chouliaraki 2006, 83), shares an equally divided media environment. I argue that this division is not purely a product of personal differences in taste but is indicative of classed moralities that underpin everyday taste judgments, following the important sociological insights of Skeggs (2004) and Sayer (2005).

This chapter additionally reflects on the debates of ecological ethics (discussed in chapter two), with a specific analysis of how people's direct experiences with the media influence their evaluation of particular practices of representation and overall operations of the 'mediated centre' (Couldry 2003). I argue here that divergent classed moralities of respectability inform their judgments of how suffering is currently represented on television and how television provides its own quasi-welfare state forms of recognition and redistribution (Fraser 1997; Fraser and Honneth 2003; Phillips 1999) as part of their claimed service to sufferers in Philippine society.

Class in Everyday Life: The Philippine Context

As mentioned in chapter one, previous studies have identified how factors such as class, age, gender and religion are significant in accounting for differences not in only media consumption practices but also aspects of morality that I am concerned with. In the Philippine context, social scientists underscore class as a significant category in social analysis (Aguilar 2003, 2005; Benedicto 2009; David 2001a[1976]; Pinches 1987, 1999; Tadiar 2004; Tolentino 2007, 2011). In many ways, this is due to dramatic income inequality; the distribution of families by socioeconomic class, as measured by the Family Income and Expenditure Survey (FIES) of National Statistics Office (NSO), is 0.1 per cent upper class, 24.2 per cent middle class and 74.7 per cent lower class (Vilora et al. 2013),[1] even in the most developed Mega Manila area (AGB Nielsen 2006).

Class in the Philippine context is typically defined and measured using market research categories of AB (upper class), C1 (upper middle class), C2 (lower middle class) and DE (lower class).[2] The anthropologist Pinches (1999) noted that categorizing individuals across these labels is common in everyday lay discourse. Scholarly work on media (Tolentino 2011) as well as consumer research in advertising firms and television networks use a broad set of class indicators that include income, education, occupation and household size/location/durability.

Industry indicators for the upper class include a monthly income of 50,001 pesos (715 pounds) and above, an undergraduate degree from an exclusive university, white-collar occupations considered highly paid and highly skilled, such as business executives or professionals, and a house located in an 'exclusive subdivision/expensive neighbourhood enclave' (McCann Erickson 2009). In local ethnographies, the upper class has been studied as the landed elite class (Aguilar 1998) who distance themselves in geographic and cultural terms from the poor who seek patronage (Kerkvliet 1990) and the middle class that challenges their dominance (Pinches 1999).

Middle-class individuals have a monthly income range of 15,001 to 50,000 pesos (214 to 714 pounds), college-level educations from state colleges (with or without a diploma), occupations in skilled and technical jobs (including nurse, call centre agent, overseas worker, small-scale businessman) and houses in 'permanent or semi-permanent conditions in mixed neighbourhoods' (McCann Erickson 2009). Pinches (1999) identifies that the middle class asserts discourses of resourcefulness and hard work in contesting the value of the spoiled and privileged upper class. Middle-class individuals are however in a precarious position (Parreñas 2001; Pingol 2001), because middle-class status could be easily lost as a result of external calamities (e.g., financial crises, natural disaster) or family tragedies (e.g., death of the breadwinner, family illness) in the absence of social safety nets and welfare state provisions.

Lower-class individuals have monthly household income levels of below 15,000 pesos (213 pounds), high school or elementary education, occupations as unskilled labours (including plumber, vendor, janitor, maid), and live in semi-permanent and temporary homes, usually in slum or squatters' communities (McCann Erickson 2009). The lower class has been subject to studies of coping mechanisms in light of poverty or disaster (Bankoff 2003; Hollnsteiner 1973; Jocano 1975) and creative uses of 'idioms of persuasion, reluctance, and pity' to draw recognition from the elite (Cannell 1999; also Kerkvliet 1990). Though 'lower class', '*masa*' and 'the poor' are sometimes used interchangeably in journalistic accounts and everyday discourse, not all lower-class individuals are officially considered poor. Government statistics set the poverty line at a monthly subsistence level of 7,821 pesos for a Filipino family of five for them to meet both food and non-food needs. Applying this convention, 27.9 per cent of the entire population in 2012 is recorded as living below the poverty line (NSCB 2013), though lower-class individuals represent around 84 per cent of the total population (AGB Nielsen 2010).

I found these categories helpful during fieldwork in identifying field sites where I could recruit participants from different classes. Class in this study is seen, however, in relational rather than static terms (Skeggs 2004; 2006) and is seen as a combination of economic, symbolic, social and cultural capitals that could be converted from one to another (Bourdieu 1986).

In succeeding chapters, 'upper class', 'middle class' and 'lower class' will be used as a shorthand to refer to my participants of certain occupations whom I recruited in specific locations across Manila. In my analysis, I refer to specific respondents without any intention to generalize about entire social classes. Of course, I recognize that the Philippine class structure does not consist of just three classes, but there also exist socio-demographic differences and sub-classes within classes (see Benedicto 2009; Pinches 1999).

Though class is the focus, I also made sure to be attentive to gender, religion and age in recruiting participants. Gender has been a significant category in studies about suffering (Melhuus 1997) and mourning (Cannell 1999). Religion, particularly Catholicism, has also been identified as a resource people draw from in making moral judgments about correct ways of coping with suffering (Bankoff 2003; Melhuus 1997). During fieldwork, it was important to listen to how Catholic beliefs about prayer, mercy and redemption were reflected in people's talk about suffering, given that the Philippines is the third largest Roman Catholic country in the world, with over 80 per cent of the population identifying as Catholics. In relation to suffering, age as an analytical category has been less prominent, though found by Dalton et al. (2008) as providing younger people (who have fewer material and political resources than older people) with moral justifications for non-action towards televised suffering.

Access to media

Regardless of class, all 92 of my respondents own at least one television set. This finding resonates with the industry statistic that television is the most pervasive medium, enjoying a reach of 98 per cent of the population (McCann Erickson 2009). The most affordable black-and-white television sets, used in low-income households, cost roughly 3,000 pesos (40 pounds), the equivalent of a housemaid's monthly salary. Given the country's proximity to electronics manufacturing hubs Taiwan and China and its own storied history of television manufacturing (Sioson-San Jaun 1999), television sets from low-end black-and-white television to high-end LCD screens are popular –even considered basic – household commodities. Cultural historians have long expressed bewilderment with (poor) people's practices of saving their wages to purchase a television 'even before a refrigerator' (Abrera 1999, 122), and media history books cite that even in the 1960s, the black-and-white television set was the top-selling appliance, 'with the electric iron a far second' (Sioson-San Juan 1999, 79).

Cable television subscription rates are lower in Manila than in rural and less developed towns. Manila residents pay an average subscription rate of 500 pesos (7 pounds) per month for a cable television subscription. Fifty-one of my 92 respondents claimed to have access to cable television and its more than seventy channels. Seven of these come from the lower class: three (housemaids and drivers) cited accessing cable at their employers' residences, whilst four admitted to obtaining illegal access by tapping into their neighbours' cable connections, a popular practice that rather belatedly became punishable by law (GMA News Online 2008). One middle-class respondent confessed that he too 'used to' have an illegal connection himself.

Industry statistics, on the one hand, indicate that average viewing times of television range from three (McCann Erickson 2009) to over seven hours a day (AGB Nielsen 2010; Burgos 2010). In contrast, television's old rival radio has dipped in listenership and primarily involves ambient rather than engaged listening (Carat 2009). Most of my respondents admitted to listening to radio only when in transit, during an energy blackout or 'when there's nothing on TV'. Industry data also note a decrease in newspaper readership in recent years (ibid.). Many of my upper- and middle-class respondents expressed that they have increasingly turned to the Internet, since online sources offer similar material as newspapers but are updated faster and can be accessed at less cost. My lower-class respondents revealed that newspapers were not significant in their everyday lives, though a few men admitted to buying the occasional tabloid for 'sexy pictures' of actresses. When it comes to cinema, only middle- and upper-class respondents reported having watched movies in the theatre in the past month. Hollywood blockbusters and local romantic comedies (produced mostly by subsidiaries of the television networks) are popular choices and are watched in multiplexes found in shopping malls. Lower-class respondents watched similar titles by purchasing pirated DVDs, which are not only cheaper (80 pence per pirated DVD versus 2.50 to 4 pounds per movie ticket) but can be watched, replayed and shared by family members and neighbours.

Traditional media of course exist alongside new media that compete for audiences' attention. Eighty-three of my 92 respondents own mobile phones. The nine without phones are either older unemployed low-income people who find phones 'unnecessary' or college students who are in the process of saving their allowances to purchase one. Whilst the Philippines is known as the 'texting capital of the world', with an average user sending six hundred text messages each month (Dimacali 2010), ownership of mobile phones in the country is currently recorded at 100 per cent (Business Mirror 2013). High mobile phone usage is said to compensate for the prohibitive costs of personal computers. Statistics agencies indicate that 36 per cent of the population have access to the Internet (Albert 2013), with Internet cafes being the primary point of access, but private Internet surfing (at their homes or through mobiles) is seen also to increase in recent years (AGB Nielsen 2010). Nevertheless, at the time when the project was completed, the Philippines was listed as eighth in the world in number of Facebook users ('Countries On' 2010) and twelfth in Twitter users (Evans 2010). At present, the Philippines has been renowned as the 'Social Networking Capital of the World' (Manila Standard Today 2013), with the 'most active' (Dumalao 2013) and 'most engaged' (ABS-CBN News Online 2010) users of social media sites.

Twenty of my respondents – all aged 30 and above – do not use computers and do not access the Internet (although the middle- and upper-class respondents potentially have access to computers either at home or at Internet cafes). These respondents are unable to operate computers by themselves and depend on their children or other members of the household to send the occasional email. Meanwhile, respondents below 30 are active users of the Internet: upper-class respondents usually log on everyday to surf websites (particularly Facebook and other social media) and download US television shows and films; lower-class respondents log on once or twice a week and maintain profiles on Facebook and Friendster.[3]

This inquiry into audiences' access to media reveals that, in spite of dramatic income inequalities present in Filipino (and in this case, Manila) society, the same media technologies and platforms are nevertheless within reach of people regardless of income. Even new media such as mobile phones and personal computers are accessible to my lower-class respondents living in slum communities: I remember here how one 20-year-old student admitted to me that she would regularly skip lunch to save up for fortnightly visits to Internet cafes.

As mentioned in chapter two, scholars such as Tolentino (2011) have remarked that 'sachet economics' applies not only to fast-moving consumer goods in the Philippines but also to media products and services. The aggressive production and distribution of mass-oriented budget versions of television sets, mobile phones, cable subscription contracts, Internet kiosks and low-denomination pay-as-you-go phone cards enable lower-class consumers to access traditional and new media, even within daily contexts of hunger and deprivation. On the one hand, this can be seen as desirable in that it democratizes the different affordances of media platforms, from mobile-phone parenting (Madianou and Miller 2011) to political learning (Pabico 1999), among the poor. On the other hand, a pessimistic reading of this relatively democratic media landscape (at least in the context of gross income inequality) is that these are market-driven practices to appease the poor and provide them fantasies that obscure the pains of poverty – for the benefit of big business and ruling elites (Tolentino 2011). Just as Tadiar (2004, 83) has said that '[the urban poor] are at once marginalized and essential' in the context of production and labour, this idea also rings true in the context of consumption. In the same way that the mobile phone market is significantly dependent on low-income, pay-as-you-go users (Madianou and Miller 2011), television ratings and advertising revenue depend on the DE market that comprise two-thirds of total viewers and consumers.

In the next section, this picture of a relatively 'democratic' media landscape is complicated once we enter the actual everyday life contexts in which media consumption is played out. In spite of access to more or less similar media, we find that respondents' actual practices with and evaluations of media significantly vary among different groups of people. Whilst inquiring about media access reveals more similarities, exploring actual practices unveils dramatic differences.

Classed Contexts of Viewing: Television as Flow or Interruption

Katipunan Avenue is a long and busy highway in Quezon City, Metro Manila. It has three lanes in most sections – four in its busiest stretch. It has one flyover, an elevated highway where motorists gain a bird's-eye view of the bungalows of Marikina Valley on one side and rows of grey cement-blocked shanties on the opposite side. The flyover connects to Katipunan's best-known landmarks: its universities. It also houses many condominiums, restaurants, supermarkets and cafes, with links to several gated neighbourhoods.

As with most parts of Manila, rich and poor are rarely geographically far from each other; though often the intersection of their paths still manages to register anxiety. Whilst a popular hangout of middle- and upper-class young people, Katipunan is also home to many street children, who are often seen begging for alms and rummaging through garbage. During my fieldwork, there was a scandalous news story about a boy of around seven who offered to masturbate a university student after the latter ignored the boy's initial request for alms (Mawis 2008). There are also fairly regular reports on hold-ups of students and business establishments (Ongpin 2006).

I came to Katipunan frequently to meet students and employees of the universities in the area, and to conduct fieldwork in the Park 7 slum community, quietly tucked away from Katipunan's university scene and unfamiliar to many of my own upper- and middle-class respondents.

Park 7 is located in the street parallel to Katipunan – a street similarly lined by condominiums, restaurants and Internet cafes frequented by students and neighbourhood residents. From the street, Park 7 fails to grab attention. Its entrance is partially obscured by trees and shrubbery; one can make out from a distance a only modest, low-ceilinged, open-air chapel. It is understandable that one's attention is instead drawn to the glass-walled, air-conditioned Pentecost Church opposite the Park 7 entrance.

Entering the narrow alleys of Park 7, the sounds of car horns and engines from the highway are immediately drowned out by the high-pitched voices of children, obviously delighted with their outdoor play, and the amplified echo of television sets all tuned in to the same channel. Houses here not only stand next to each other in close proximity but even lean against each other for support. Open windows give not a glimpse but the full view of people's living quarters, usually ten-by-ten feet and illuminated by light from the television. Many households own individual television sets, breaking away from the stereotype of slum dwellers huddling together with necks craned outside a neighbour's window to catch their favourite programme (e.g., Abrera 1999; Burgos 2010). Houses here, at least those with residents presently indoors, seem to have their television sets on at all times.

At Park 7 any discussion about media automatically begins with people expressing monikers of self-identification with either of the two rival networks: *Kapamilya* (Of one family) for ABS-CBN viewers or *Kapuso* (Of one heart) for GMA Network loyalists. Self-identifying as *Kapamilya* or *Kapuso* means devotedly tuning in to one channel and being familiar with its programmes, celebrities, journalists and public service initiatives. I learned through repeated visits that most residents in this particular community are *Kapamilya*, whilst residents in a nearby slum community – 'those across the creek', in their words – are *Kapuso*.[4] Self-identifying *Kapamilya* respondents here either have personal experiences of visiting ABS-CBN studios or know neighbours in the community who have done so. Although my Park 7 respondents confess to occasionally switching channels to the rival network for a specific programme or two,[5] frequently the television is fixed to ABS-CBN Channel 2.

To illustrate: part-time laundrywoman and seamstress Norma visits her employers thrice a week and spends half a day doing their laundry for a monthly pay of 4,000 pesos (60 pounds). When she returns in the afternoon, she immediately switches on the TV for her afternoon soaps whilst doing household chores and/or clothing repairs for her other clients. When her 11-year-old son arrives from school at five o'clock, she begins preparing dinner, and her son watches the late afternoon cartoons (on the same channel). Her husband comes home from his job at a shoe factory at around seven o'clock. They watch the news and evening soaps altogether. They switch off the TV only when their son needs help with math homework. On days when Norma has no work, she says she has more free time for television. She would tune in to the noontime show *Wowowee* (a programme I will return to in chapter four), often in the company of her next-door neighbour Elsie.

Watching television is frequently regarded here as a *libangan* (enjoyable activity; something with which to pass the time). As a constant domestic companion, it is said to 'enhance' or 'spice up' mundane everyday life: the word used is *dagdag-libangan*; literally, 'adding enjoyment or leisure'. However, the word *libangan* also has a connotation of passivity in that it constructs television consumption as an activity that keeps a person occupied in the absence of worthwhile or urgent tasks. After all, its converse *walang mapaglilibangan* means 'having nothing to keep one occupied'. This ties in with respondents' descriptions of their mundane daily activities.

Some older people, especially those who were undergoing (or have undergone) particularly difficult struggles, express the belief that television 'makes the day go by faster'. I remember here the 51-year-old fish vendor Ofelia. Ofelia moved to Park 7 a few years ago after a typhoon caused floods to wash away their squatters' settlement. Her family had been subsisting on a combined monthly income of 5000 pesos (70 pounds) from her work in the Pasig fish market and her husband's job as a carpenter. She narrated the difficulties of relocation to Park 7 and the unstable salaries that both she and her husband had earned since moving to Katipunan. Her 23-year-old daughter was also unable to finish college and was now jobless and pregnant with her second child. Ofelia shared,

> Television helps me get through my day, my child [*anak*]. There are really difficult days sometimes: when I get up in the morning sometimes I have this cold sweat because, you know, you know that you'd need to work extra hard and pray extra hard that [all] goes well. I don't work every day so I just wait for my husband, or my children. And instead of, you know, sitting in a corner worrying or crying about life, at least there's the TV. It entertains you, it inspires you. And you watch other people and you hear their stories. 'Wow,' you say, 'I've been spared. I'm actually pretty lucky compared to them.'

Local commentators often claim that television 'peddles fantasies' and 'escapism' to the poor (Coronel 2006a; Inquirer.net 2006). However, Ofelia, along with many low-income women I met, does not and could not escape from realities of hardship and suffering in the sense that television transports them to alternative realities of comfort and luxury, as they do in India, where television is charged with being 'in denial' of suffering (Mankekar 1999; Sainath 2009). The soap operas they follow feature the popular historical archetype of the poor *atsay* (maid) who suffers at the hands of her wealthy employer (Flores 2001); the news they watch features crimes, fires and police confrontations in squatters' communities; and talent shows are as much about contestants' emotional backstories of separation, poverty and personal tragedy as they are about actual displays of talent. Television, as both object and text, helps manage the conduct of their everyday lives by providing a constant and reliable frame to enhance its mundane domesticity and, consistent with Silverstone's (1994; 2005) arguments, can provide symbolic resources for hoping and imagining stability in contexts of anxiety and uncertainty. I discuss further this sense of 'ontological security' (Giddens 1991; Silverstone 1994, 2005) that some audiences derive from television, and especially the news genre, in chapter five.

Many people in Park 7, I found, are underemployed and work less than five days a week. Staying at home is the most practical option for people with little income who are in the process of paying off debts to employers or relatives. Nilo, a 35-year-old male whom I always saw watching television in a *sari-sari* store,[6] puts it this way, 'People always say that we should go out to the city, take risks, find odd jobs, look for employers. Don't they know we've tried that and failed? It's either we end up spending money or getting into a fight. And our wives will just think we're looking for other women, you know?' Television then becomes the obvious and least costly *libangan* that 'makes time fly' for people made immobile by poverty, whilst offering hope through media practices of representing other poor people and retelling their stories in both factual and fictional genres.

In contrast to low-income respondents, upper-class and most middle-class respondents spend less time at home and are thus less likely to watch television at length, especially during weekdays. Young professionals below 30 and senior executives above 40 tellingly use the phrase 'lazing around' to describe the 'luxury' of watching television all day that they rarely experience; they say that even their weekends are increasingly spent outside of the home. This resonates with market research that describes the mobile lives of the middle and upper classes (McCann Erickson 2009) and their metropolitan experiences of 'flows' and 'fluidities' (Tadiar 2004).

Television for them is not often experienced as part of the natural rhythm of the day but as a treat, punctuation or interruption. As a treat, working professionals switch on the television to reward themselves with leisure time at the end of a working day; HBO, ETC and AXN are some of the favourite channels.[7] As punctuation, television is experienced as discrete, particular programmes, especially for younger people who illegally download American television shows such as *Desperate Housewives*, *Grey's Anatomy* and *Lost*. Downloading content enables them to catch up with particular shows at their leisure, rather than follow the schedules of cable television channels. As interruption, Philippine television programmes only seem to command attention in times of local public crises, natural disaster or political or celebrity scandal: from friends' alerts usually via text message, upper-class people tune in to the top channels they otherwise would not surf through ('I only surf from channels 17 onwards,' says Candy, referring to her practice of skipping local terrestrial Channels 2, 4, 5, 7, 9 and 13). In my interviews with upper-class respondents, they often appeared hesitant and embarrassed to talk about occasions when they do actually watch ABS-CBN or GMA. I remember here an interview with Patricia, a businesswoman in her forties, a volunteer for charity organization *Gawad Kalinga* (Bestowing care), and mother of two students at a private university.

JONATHAN: So aside from watching cable for about two hours every night, do you also watch ABS or GMA?

PATRICIA: Oh! Haha! Do I? Oh God, very rarely. I don't even remember the last time [...] Ahh [...]

JONATHAN: What about [the Sunday gossip shows]? Or the news?

PATRICIA: Oh yes, the news [...] What was the last I watched? Oh I know, this is funny, but I remember watching the news and gossip shows for a while during the [sex video scandal].[8] You know, you have to be updated! I'm sure you watched it too! Oh the drama! Haha!

Patricia here, just like many upper-class respondents, does not display much knowledge about local television and its programmes and personalities. Respondents like Patricia would be hard-pressed to name the titles of soap operas currently airing on prime time. Though they can often recall names of Filipino celebrities, they do so in reference to billboard advertisements visible along major highways, and not from direct consumption of their films or TV shows.

The upper-class people are often self-aware, even eager to display their ignorance of Philippine popular culture. In one interview, a young professional began by saying, 'Hey, I hope this interview is not a quiz about local showbiz, okay?' The research encounter in itself proved to be a moment when class is performed. Upper- and (some) middle-class respondents were keen to display reflexivity about academic criticism of local television as well as the popular discourse of trashy, low-culture, *masa* (mass) television. Conveying ignorance about local television can be seen here as a strategic performance where the speaker perceives that she accrues for herself higher value in dialogue with me as a researcher.

Upper- and middle-class respondents seemed to feel a need to provide justifications for tuning in to ABS-CBN or GMA. Social pressure, especially in relation to the need to keep up with celebrity and political scandals, is one oft-cited justification. For some, familiarity with local television, especially the news, is described as a civic, even patriotic, duty – consistent with research in other countries (Jensen 1995; Putnam 2000). Private university students and charity group members claim that seeking information about public issues is a way by which they can become good citizens by being able to address these social problems 'in their own humble ways' perhaps in the future. For the most part however, geographical divides between zones of safety and zones of danger are replicated and reproduced in symbolic spaces of media, themselves splintered along class lines.

In the next sections, I explore in greater detail the reasons for and motivations behind class-divergent practices of 'switching off' and tuning in to local television.

'Switching Off': Disgust with the *Jologs*

In the political communication literature, 'switching off', particularly from the news, is an indicator of citizens' political and civic disengagement (Couldry et al. 2007; Putnam 2000). In media and migration studies, it can be taken as evidence of social exclusion (Gillespie 1995; Madianou 2005b). And in the media ethics literature, it is normatively denounced as an example of 'active denial' (Cohen 2001; Seu 2003), presumably by privileged (Western) audiences, in purposefully avoiding negative feelings of guilt or responsibility for distant others. Silverstone (2007, 131) describes this 'capacity and desire to ignore and forget the reality and pain of the other's everyday life' as a form of 'collusion'.

'Switching off' in the upper-class Filipino context sounds different from these existing Western descriptions at first. My respondents often describe their choice of television programmes and favourite media personalities as a matter of 'taste' and 'personal preference'. Here is Jake, a 21-year-old university student.

JONATHAN: So how come you don't watch ABS or GMA?

JAKE: Oh that's just me! I really don't know. I guess it's just a personal [...] thing. You know? We choose different cars or pick different colours. I just never developed the taste for it.

JONATHAN: So at home your family doesn't watch it either?

JAKE: Yeah. Umm, my mom does. Sometimes she watches. During Sundays, she watches. But I guess that's a girl thing.

JONATHAN: What do you mean?

JAKE: Oh you know. Because it's talk shows. Girls love that sort of thing. I don't. Most boys don't.

(Individual interview -upper class)

Especially in my interviews with students and young people below thirty, I was struck by how non-consumption of local television was taken as natural, with little need to explain or justify this choice. 'Taste' and 'personal preference' – or in Jake's case, a belated attribution to gender as an influence on personal preference – are often mobilized by respondents as self-evident explanations for 'switching off' from local television.

There are many others (including some middle-class respondents) who explain their preferences by referencing the 'superior' and 'original' qualities of Western media productions.

BENJIE: Oh my god. Our programmes are pretty pitiful. Our production values seem stuck like they're in the '80s. Even our news! I get dizzy trying to read the [news] ticker in [GMA's] 24 *Oras*. I think it's a combination of the font, the colour and maybe the resolution of the graphics. It makes me dizzy.

(Individual interview – middle class)

PEPPER: I find [local TV] too [...] crude. And backward. Like, when we import reality shows, they often botch it completely.

JONATHAN: Can you give an example?

PEPPER: Yeah. Look at *Pinoy Big Brother*. The entertainment of the original is all lost. Here they ask the viewers to decide who to evict every week. In the US, the housemates vote each other off. [The US version] is more entertaining, right? Because there's strategy and conniving. Here, it's all obvious. Of course whoever's the poorest will get pity votes from the *masa*.

JONATHAN: So you mean, [*Pinoy Big Brother*] is more predictable?

PEPPER: Yeah it's predictable because it's for the *masa*.

(Group interview – private university)

Academics and cultural commentators have long remarked about Filipinos' 'colonial mentality' and 'white love', particularly for American goods and products (Rafael 2000; Tadiar 2004; Tolentino 2011). Political economy scholars have also drawn links between the period of American colonialism (1898–1946) and the contemporary American imperialist practices of dumping its cultural slop on, and profiting from, its one-time colony (Bello 1999). There are surely some traces of 'white love' (Rafael 2000) here in audiences' appraisal of 'grand' and 'innovative' American imports in comparison to lower-budget and 'derivative' Filipino productions and adaptations. But in this context, similar to Daniel Miller's (1992) findings in the context of Trinidad, access to and praise for American programming are less indicative of a desire to become American than they are revelatory of local strategies to accrue higher value for oneself within local class hierarchies.

In the exchange with Pepper above, we hear telling mentions of 'the poor' and 'the *masa*'. In this context, 'the poor' and 'the *masa*' are used not only as explanations for the lower value that local productions have compared to international productions, but also as justifications for upper-class audiences to 'switch off'. Local television, being oriented to low-income groups, is assumed to always-already follow generic conventions that appeal primarily to their tastes.

In the case of the well-known reality show *Big Brother*, Pepper laments the localization process of the US programme that she was familiar with[9] and how (in her view) the original version's individualistic and entertaining spirit has been lost in the Philippine version. In the local version, 'Big Brother' is referred to by the contestants in its literal Tagalog translation *Kuya* and acts as a benevolent, though at times naughty, older brother. His disembodied voice instructs houseguests to perform household chores associated with *atsays* (maids), interviews contestants about their personal tragedies à la *Wowowee* and rewards winners in cash and kind as the networks' charities do. Big Brother, or *Kuya*, functions here as kin, literally an extension and (dis) embodiment of *Kapamilya* (ABS-CBN's tagline), by referencing the network's other genres and social services. Though the selection of *Big Brother* housemates often represents a cross-section of society as it does in international versions of the show (Klein and Wardle 2008), the few upper-class housemates are often eliminated earlier than the majority lower-class housemates as a function of audiences' participation in saving rather than evicting contestants – a mechanism that deviates from other versions of the show.[10]

This may explain why respondents immediately assume that any one programme is automatically metonymic of the television network and the values of service and entertainment for the poor that it claims to uphold. There is a perception among the upper class that the poor are over-represented in the sense that a disproportionately high number of poor people are on television, with the programming excluding the middle and upper classes. However, they nevertheless deconstruct media's acquiescence to the poor as a strategic move on the part of media to seek out ratings and profits. And they perceive the poor as situated in a highly asymmetrical, even abusive, relationship with these powerful and instrumental institutions. I return to this in succeeding chapters.

Over the twenty months of my fieldwork, I also often heard upper- and middle-class respondents describe local television as *jologs*. *Jologs* shares connotations with the equally classist British insult 'chav' in referring to cultural objects, geographical locations and ultimately people judged as lacking in taste, respectability and value.[11] A Filipino blogger's definition of *jologs*, at least as it pertains to cultural objects, goes,

> I don't really know how to define *jologs*, but I will try to give an example on how something can achieve *jologs* status. Say a certain footwear became the 'in' thing in the Philippines. This footwear is really expensive for just a pair of rubber stuff being worn [sic] on the smelliest part of the human body. But since the Philippines has the creativity to pirate anything fast, the said footwear now costs only 50 pesos (80 pence) and is readily available at your local [market]. Now, everyone from your driver, to your laundrywoman's son, owns the identical footwear you are wearing. The footwear is now 'mainstream' in the sense that the 'representative population' of the Philippines, a.k.a. 'the masses', can now own and wear the footwear. Congratulations, your footwear is now *jologs* (Uy 2009).

Academic references to the *jologs* (Benedicto 2009) and related discussions of the 'new rich' (Pinches 1999) have largely associated these terms with class contests between the high-cultured upper classes and the formerly poor but now upwardly mobile groups of people in Philippine society, such as middle-class overseas Filipino workers, middle-class nurses and upper-middle-class (but 'uncultured') Filipino-Chinese businessmen. My fieldwork discovers, however, that *jologs* is used as a derogatory term by upper- and middle-class viewers towards the space of television and its *masa* inhabitants, or the 'representative population of the Philippines', in Uy's (2009) words above.

JONATHAN: So how are TV shows *jologs*?

TAMARA: Oh lots of things. The noise, the songs, the loud music in a variety show. The plot can be *jologs* too.

JONATHAN: How can a plot be *jologs*?

TAMARA: [In a soap opera] Such as when it's the stereotypical maid being abused by the rich employer [storyline]. And then later she's saved by this guy, who's usually not even that handsome! And yeah, all the dramatic moments: the close-ups, the crying, the hair-pulling, the cheesy romantic lines. Oh and the love songs too! *Jologs*!

JONATHAN: And the actors are *jologs*?

TAMARA: Oh yes. Everything *jologs*. Like, have you heard them be interviewed? Some of them may look good, but when they open their mouths, they sound like they're from the *wet market*. TV is like one big *palengke (noisy wet market)*. And literally some of the celebrities may have once been [fish vendors] themselves!

(Individual interview -upper class)

Television programmes are judged as *jologs* for their aural and visual vibrancy and excess. Clichéd plot points and motifs of *sigawan*, *sampalan* and *sabunutan* (shouting, slapping and hair pulling) that my respondents associate with lower-class interactions in public wet markets also confer *jologs*-ness to a television programme. Crucially, television celebrities and even journalists can be *jologs* themselves: Journalists are made fun of for their (rare and inconsistently successful) attempts to speak English on television, and local celebrities are also ridiculed as having 'no breeding' – judged from the way they wear cheap and/or skimpy outfits on noontime shows to their eagerness to disclose intimate details of their personal lives on talk shows.

Like chavs, the *jologs* are judged as lacking the 'right' sort of cultural capital by having too few resources that confer value on them (such as being unable to afford the 'right' clothes or being unable to speak English fluently) or being too excessive in their conduct of the self (such as overaccessorizing one's body or overtly displaying emotions in public). Lack and excess however are crucially intertwined, as Skeggs (2004) and Tadiar (2004) have astutely pointed out.

In this context however, *jologs* seems to cast a wider net than chav by being a totalizing description that derides not only clothing styles and accents, but also even genre conventions, television narratives and television channels themselves ('everything *jologs*', as Tamara says). *Jologs* identifies cultural objects, properties and embodiments of the lower class in Filipino society. *Jologs* attacks the cultural objects produced for their consumption. It attacks the televised public space that is perceived to over-represent (Wood and Skeggs 2009, 177) the poor and exclude the rich. It attacks the presence of 'noisy' and 'undesirable' people both within and without the frames of the living room television. It is ultimately a 'word of hatred'[12] for the bodily existence of the lower class in a metropolis historically defined by the perpetual erection and transgression of borders of class difference.

Using *jologs* to describe television can be interpreted as a peculiar and local form of 'active denial' (Cohen 2001; Seu 2003) and 'collusion' (Silverstone 2007). However, turning away from television in this case is not always a successful project of 'forget[ting] the reality and pain of the other's everyday life', as Silverstone (2007, 131) has described in the Western case. Pain and suffering, after all, are visceral realities that no Filipino can truly honestly forget in a metropolis with parallel yet also intersecting zones of safety and zones of danger, in the same way that pain and suffering are guaranteed to be present 24/7 in one's living room through the mediations of local television.

In the Filipino upper-class case, the use of *jologs* first and foremost suggests that sufferers are not successfully denied, but implicitly *recognized*; *jologs* suggests a hostile acknowledgment of the presence of less privileged others in Filipino public life – others who are readily readable in and out of the television through their cultural products, social practices, forms of speech and actual physical bodies. Using *jologs* is not (just) a matter of avoidance or looking away in the sense of apathy or disinterest, but it suggests a more active form of sanctioning that further marginalizes these less privileged and yet always proximal others. *Jologs* actively attaches lower value to objects produced and consumed by 'the representative population of the Philippines' and ultimately to their very lives. *Jologs*, rather than simply being about taste and preference, reveals the deep-seated upper-class anxieties and fantasies of their project of fashioning desirable selves amidst an environment of poverty that exists too much in excess (Tadiar 2004, 92). 'Switching off' from local television as a space of the *jologs* can then be viewed in parallel with observations of the elite's geographical distancing from less privileged others: from practices of 'flying over' in the metropolis (Tadiar 2004) to 'living far apart' from the poor (Kerkvliet 1990). The elite's obligations to 'those who have nothing' (Cannell 1999) are actively reduced through the act of branding less privileged others as having low value, in this case, low intelligence, culture and moral worth.

A form of 'switching off' was also observed among a few (usually male) lower-class respondents. Gender differences were evident in how males complain about their wives specifically – or all women generally – for either being melodramatic in their insatiable appetite for stories of triumph over tragedy on television, or being lazy by staying at home and 'doing nothing'. Some low-income male respondents associated excessive television consumption with a lack of potency and resourcefulness. However, I suspect that this also had as much to do with the high-emotion content of local television as with the physical location of television consumption within the domestic sphere of women. In this light, men staying at home run the risk of being devalued and branded by their peers as feminine. This is perhaps why I observed that many low-income males watch television with other males in public places, such as *sari-sari* (variety) stores, canteens and beer gardens.

'I'm Just Surfing Through': Re-Brandings of the *Jologs*

In spite of the upper-class branding of local television and its viewers as *jologs*, I also found during my fieldwork instances where upper- and middle-class respondents also tuned in to television and were familiar with local programmes, celebrities and journalists. This section sketches a more complicated picture of consumption – particularly, more middle-class consumption – whereby privileged people in Philippine society may perhaps occupy a more morally desirable mediated orientation which at least involves basic 'awareness of events around us and of the people who make up our society and wider world' (Ellis 2009, 83). However, I argue here that affinity and familiarity with local television can also be used not only and primarily to gain empathetic understanding of the other but also as a personal strategy to gain positive value for oneself in contexts of intense class conflict and position-taking. Nevertheless this accrual of positive value is never stable and always contested; it is a strategy that not all people can mobilize, as its successful deployment ultimately depends on harnessing cultural and symbolic capitals exclusive to certain classes.

As mentioned above, television consumption can function as treats, punctuations and interruptions in the everyday lives of middle- and upper-class Filipinos. Their mobile lives and busy schedules do not lend themselves to extended stays at home, and therefore television is rarely experienced as a constant flow and ever-present backdrop to their everyday lives. However I also met middle- and upper-class respondents who claimed to be regular viewers of select local programmes, such as news broadcasts, soap operas, talk shows and noontime shows. Some confessed to watching local television when they chance upon something interesting when they are 'just surfing through' (*napadaan lang*).

JONATHAN: So you mention that you sometimes watch local TV, especially on Sundays? Can you describe the last time that happened?

PAUL: Oh okay. Yeah, as I said, it doesn't really happen very, very often. Like last time, I just sat on the couch and flipped the channel to watch Lost [on cable TV]. It wasn't a very good episode though. It's a rerun and I've seen it before [...] And so flip-flip-flip with the remote [...]

JONATHAN: What channels do you flip through?

PAUL: Well, I surf through them all. And I was just surfing through the local channels, and then I saw [actress] Angel Locsin on [the noontime show]. I haven't really seen her dance before. So that interested me and so I kept it on there for a bit.

JONATHAN: But you know her from her soap operas?

PAUL: Yeah, she's very pretty in the soaps. Yeah, I know, I know. That's kinda *jologs*. [laughs] So only when I'm surfing through. But don't get me wrong, I enjoy [local television] too. I can be slightly *jologs*. [laughs]

(Individual interview – middle-class)

This excerpt illustrates a case where class is reflexively performed in the context of the research interview. Paul displays reflexivity about the 'low' value of local television and justifies here why he even tunes in: the absence of alternative foreign programming is used as a justification, as is his accidental manner of encountering a rare scene on local television. By saying that he was 'just surfing through' and at the same time admitting to his 'slight' jologs-ness, advertising manager Paul nevertheless distinguishes himself from low-income viewers typically stereotyped by middle- and upper-class audiences to be addicted and loyal *Kapuso* or *Kapamilya* – the *true jologs*.

Some of my middle-class respondents were less apologetic than Paul, however, in reflexively and humorously self-identifying as *jologs* themselves, sometimes even at the very beginning of the interview.

JONATHAN: So what are your favourite television programmes?

AMANDA: Oh. Gosh, let's be clear about this. You'll call me *jologs*, but yes, I enjoy GMA shows. I think they're good quality and down-to-earth and funny! Like, did you watch [the episode of] *Bubble Gang* last week?

JONATHAN: Oh no! I didn't. I was [...]

AMANDA: Oh my. You should. You should. You should so you could relate with the *masa* (masses). The problem is not all people watch local TV that's why they can't relate to what's going on.

(Individual interview – middle class)

Amanda is a 31-year-old employee of *Gawad Kalinga*, a Catholic charity organization that specializes in building houses for the poor. Amanda is not like other *Gawad Kalinga* donors and volunteers, typically upper-class executives residing in the gated neighbourhoods of Manila. As a mid-level employee, she earns around 10,000 pesos (140 pounds) a month and still lives in her parents' house.

I suspect that, at the very beginning of our interview, Amanda assumed that I had come from an elite background and as an academic would negatively judge her taste in television (as many academic and journalistic pieces on local entertainment do). Amanda then asserted *jologs*-ness (exemplified by her consumption of local television) as a positive trait that she explained as enabling her to keep in touch with reality and to relate with the masses – in opposition to the Filipino upper class. As the interview went on, Amanda would tell me about her own experiences in the charity organization: how upper-class donors expected houses to be built instantly and how they expected the poor to dutifully follow the rules of their charity organization, such as pleasantly greeting visitors or volunteers, attending Catholic mass regularly and keeping their houses clean and spotless. I suspect then that Amanda's middle-class consciousness was particularly fertile because of her diverse and messy experiences of interclass interactions, and that may explain her positive re-branding of *jologs* when she introduced herself to me (as a perceived highbrow academic).

I also remember another instance when being *jologs* and *masa* was re-branded positively by respondents and even popular figures of Manila's socialite scene. For instance, I remember the underwear fashion show hosted by top local clothing retailer Bench that became the 'it' event across multiple sectors of Manila society in July 2009.[13] Television fans were excited to see their favourite actors and actresses wearing only skimpy Bench underwear. To obtain (bleacher) tickets, fans had to purchase upwards of 500 pesos (7 pounds) worth of Bench merchandise over a period of several months. The middle and upper classes, in addition to purchasing merchandise in-store, also made use of personal connections to industry and media insiders to score better seats in the fashion show. Specifically, female respondents aged 30 and above, from both the upper and middle classes, confessed to the guilty pleasure of seeing well-built and scantily clad men: the very men typically derided in their everyday talk as *jologs* television actors.

In the case of my female middle- and upper-class respondents, there appeared to be some form of fetishism with the lower-class male body. Some of my respondents singled out, for instance, GMA actor Aljur Abrenica for having 'the body of a carpenter' and '[looking] like [a] gardener' and expressed desire to ogle his figure in the underwear show. Some older middle- and upper-class women expressed a preference for Aljur Abrenica's bulkier, 'very manly' body (*lalaking-lalaki*, also translates: a true male) over the leaner, gym- or yoga-toned look of other – more well-off – actors and models. In their case, Aljur Abrenica, whilst being read as *jologs* for his darker skin, bigger body and profession as television actor, was re-branded as more desirable than less *jologs* celebrities because of his raw sexuality and true masculinity.

At the same time, I could not but suspect that this increase in Aljur Abrenica's desirability among upper- and middle-class women was enabled by his appearance as an underwear model disrobed of almost all clothing. Stripped of signifiers that may connote *jologs*-ness, the naked and silent Abrenica becomes a more acceptable object of desire than the clothed and voluble Abrenica who conveys *jologs*-ness through appearance and speech in soap operas and talk shows.

In these examples above, the re-branding of *jologs* from having negative to positive value is done in both moral (in the case of Amanda) and gendered/sexualized terms (in the case of the female fans of Aljur Abrenica). Being *jologs* – or at least 'slightly *jologs*' – is one way in which the middle-(and some upper-) class participants contested the merits of the impermeable boundaries of Filipino sociality. Whilst upper-class high society is highly respected for their cultural and symbolic capital accumulated over generations (McCoy 2009; Pinches 1999), they can also be ridiculed and challenged not only for the folly of boundary-raising practices amidst the liquid mess of the metropolis (Tadiar 1996), but for the moral deficiency of such practices. My middle-class respondents, like the 'new rich' contesting the 'old rich' in Pinches' (1999) study, positioned themselves as having moral ascendancy over the upper classes whom they perceived as content in 'flying over' from one zone of safety to the next, at all times in 'active denial' (Seu 2003) of the poor. Middle-class respondents asserted moral discourses of authenticity (by being connected with reality) and empathy (by being able to relate with the poor) as strategies to differentiate themselves from the 'less moral' upper class.

Additionally, I observed that *jologs* is a gendered term in that it is more acceptable and desirable when read of males rather than females. Being *jologs* has connotations of being 'raw' and 'truly male' (*lalaking lalaki*) when *jologs*-ness is seen in male bodies. Viewing *jologs* male bodies becomes sexually attractive to some middle- and upper-class females, who may be resisting traditional and religiously informed Filipino standards of femininity (Hollnsteiner 1981; Pingol 2001) as well as the rigid class moralities that similarly regulate female sexuality. *Jologs*-ness in the property of females is judged more harshly in Philippine society.[14]

Reclaiming Value

It is crucial to note is that the strategic and temporary re-brandings of *jologs* as having positive value are resources less available to lower-class respondents. The poor remain fixed and essentialized as *jologs* and *masa* and cannot escape this low-valued social positioning. I remember here Emily, a 19-year-old public school student who told me that she was once called out as *jologs* and *jejemon*[15] by other users on Facebook:

> What can we do? We can't really deny that we're *jejemon*. I don't spell [words] in English that well. Some of my friends deny that we're *jejemon*. But really, who are we kidding? I say to my friends, 'Let's just be true to ourselves', you know? We text this way. We don't have shiny hair. We don't own the [latest] gadgets. It's obvious to all. So let's just admit it and be happy, I say!

Emily, on her second year in nursing school, came across as smart and articulate during the interview. But here, she ultimately resigned herself to the social position conferred on her and her peers by society's cultural and economic elites. The only way she could recuperate value for herself was to assert a discourse of authenticity, being true to oneself, vis-à-vis her other peers who more actively resisted the *jologs/jejemon* label but ultimately failed in their project. *Jologs*-ness, *jejemon*-ness, *masa*-ness, poor-ness and suffering, after all, are quite easily read onto people in a country where the material and cultural resources to transcend these are severely limited to, and vigilantly guarded by, a few. Instead of denying their situation then, their only recourse is to profess this identity. Indeed, Emily's statement above reflects the idea that, when an identity is under attack, the only viable strategy is to embrace this very identity (Arendt 1968, 17). And though it does not fully recuperate value for Emily when she admits to her *jejemon*-ness, in the context of personal testimonies of suffering on television, sometimes claiming a position of low value can productively claim compassion and recognition, as we will see in the next chapter.

For the most part, my low-income respondents value television network practices of providing television programmes and social services that address their conditions. Low-income respondents who had access to foreign programming still tuned in to local programming for reasons of relevance and authenticity. News about international politics and reality shows about celebrities' opulent lifestyles have a sharp disconnection from their everyday lives, given that they have little sense of political efficacy, or interest in luxury products or services. Though I had low-income respondents who complained that they would also feel 'sad' and 'depressed' by local news items, they nevertheless expressed a positive evaluation of local media's focus on the conditions of the poor. In their eyes, representing poverty has positive value as it creates awareness of, and possible resolutions to, the difficult life conditions of others like them. They contrast this with practices of international media and local cable channels that grant minimal visibility to poor people. I return to this in chapter five where I also discuss their preferences for gruesome rather than sanitized images of poverty and suffering.

In summary, consuming local television is a generally positive experience for the poor in spite of its association with *jologs*-ness. Unlike middle-class respondents, they cannot assert a reflexive or ironic moral discourse of authenticity that could reclaim value against the upper class; for them, *jologs* is a reality that is intimately tied to their lack of economic and cultural resources to access and appreciate higher forms of culture.

Unlike chavs in the British context, the lower class cannot assert *jologs* as a term of self-ascription that makes them socially desirable (Brewis and Jack 2010). In the context of television consumption, lower-class viewers can only admit that being *jologs* is instrumental to access the symbolic narratives and forms of recognition and reward that media offer to the Filipino poor. As Nancy, a housemaid in an exclusive subdivision, told me,

> My bosses often make fun of me when they catch me watching the local soaps. They call me, '*Jologs, jologs!*' and switch the channel immediately. I don't think there's anything wrong with what people like me watch. What we see on television is our lives. So [...] that probably just means that the rich just can't relate to our issues [...] That's why they don't think [television] is any good. But the truth is, we live in different worlds.

Direct Experiences with the Media: Media Pilgrimages and Reverse Pilgrimages

Following Couldry (2000; 2003) and Madianou (2005a; 2005b), I also explored audiences' direct experiences with the media: visiting television networks, interacting with celebrities, being interviewed by journalists and working alongside media professionals. My fieldwork found that these direct experiences with media informed how people from different classes evaluate the ecological ethics of television and in turn either tune in or 'switch off' from television. For some middle- and upper-class respondents, direct experiences with media enabled practices of charity and volunteering during times of crisis and offered opportunities for audiences to become active participants by aiding vulnerable others, as succeeding chapters will develop.

It must be noted here that the services that television networks extend to the poor do not simply remain on the level of textual representation in the sense of providing narratives about the lives of the Filipino poor in different programmes or genres. Rather, hospitality is more literally evident in how television networks are actual venues that many poor people travel to and visit in order to avail of live entertainment and/or various social services.

Outside the heavily patrolled television studios, there are almost always around fifty to one hundred people queuing. There is one queue for general inquiries with the network charity; another one or two for specific services provided by the network charity; one or two queues for studio audiences of the daily noontime shows; a queue for low-to mid-level employees; a visitors queue for the networks' suppliers, partners and employees' friends; and a special entrance for celebrities, executives and VIPs. Security guards are variably cordial depending on these queues, reserving their 'sirs' and 'ma'ams' for people in the latter queues.

As mentioned in chapter two and to be developed in chapter five, charity operations are institutionalized within the television networks and operate in synergy with news organizations. Representatives from the charity organization travel alongside broadcast journalists to cover a natural disaster; after the journalist delivers a report, they stand side-by-side her in handing out relief goods to victims. 'Reverse pilgrimages' (Couldry 2003, 93) involve media people visiting ordinary people during both exceptional events of disaster as well as in their everyday life in the city. Man-on-the-street interviews in the news often feature variety store vendors and marketplace scenes rather than wealthy urbanites in posh shopping malls.

Quasi-welfare state services are available within this geographic mediated center all year round, and not only for victims of natural disaster. Each network's charity has one manager dedicated to receiving visitors, hearing their stories, assessing their requests for aid and either providing direct assistance or forwarding them to appropriate government agencies or non-government organizations.

Among my Park 7 respondents, media practices of visiting the poor in their everyday lives and being hospitable to them in the media centre are desirable and enhance their affinity to the television networks. Not only are they informed and entertained by TV programmes, but they also feel a sense of comfort that an important social institution recognizes the less privileged and demonstrates willingness to help them. Two Park 7 respondents recall actually visiting ABS-CBN's *Bantay Bata* (Child watch): one reported a local crime in the community that had been unresolved by policemen and another requested money for a tetanus shot for her son after her son had been bitten by a stray dog. Whilst the former reported a positive experience (the ABS-CBN representative called appropriate authorities to act on the situation), the latter mentioned that she was referred to a government hospital that provides cheap tetanus shots. Those who have not visited charities also claim that it is reassuring that they have somewhere to run to in times of need (*may matatakbuhan*). I return to these themes in chapter five, where I discuss in greater detail how people's responses to the news are informed by such extra-textual knowledge of and direct experiences with media.

For my upper- and middle-class respondents, visits to the television networks are rare occurrences. For instance, the university students talked about visiting the networks as part of school field trips. The retellings of their experiences, however, do not convey much reverence in their crossing from 'ordinary' to 'media' worlds, as previously observed by Couldry (2000, 84-5) among theme park and television studio visitors. Instead, the retellings were highly intellectualized: they discussed what they had learned about media procedures and equipment, but with critical detachment about the cultural objects produced within media centres. Photographs taken with celebrities and journalists are recalled in more ambiguous terms: some registering genuine delight and others registering (again) guilty pleasure. Though several students expressed intentions of working in the media industry, they specifically identified working in networks' news organizations and not the entertainment department. News people are regarded as more esteemed and respectable, in contrast to the 'truly *jologs*' entertainment producers and writers.

Several upper- and middle-class respondents also cited appearing on television themselves. Distinctions are made between news and entertainment genres in terms of respectability. For example, one businesswoman recalled bringing her friends visiting from the US to watch *Wowowee*. She said that these 'kinda *jologs*' friends of hers pleaded that she get them tickets to the show and that she had to use her contacts at ABS-CBN to secure seats (instead of queuing like everyone else). She retells this story with cringing affect and dramatically describes how shamed she was when the camera panned to them in the audience. 'My God, I prayed and prayed that nobody [among my friends] saw me on TV.' A more positive experience with a television appearance was recounted by a banking executive who felt proud to have guested on a public affairs show to discuss the economic crisis: 'I had to make an effort to speak in Tagalog but it was nonetheless a success.'[16] For him and for other upper- and middle-class respondents, public visibility and fame accorded by television appearances may offer symbolic capital, and they are actively sought. Television appearances confer either positive or negative value based on genre and on professional motivations of middle- and upper-class people.

During Typhoon Ondoy, ABS-CBN and GMA Network, in De Quiros' (2009a) words, 'acted as the government' for spearheading and orchestrating relief efforts. In scenarios of exceptional disaster such as Ondoy, visits to television networks become opportunities for middle- and upper-class people to provide aid and assistance to the many flood victims. I return to this in chapter five.

What this section highlights is that direct experiences with the media are similarly inscribed in classed configurations in the same way that everyday practices of television consumption are. The media space in the Philippines operates to confer both symbolic recognition and material reward or redistribution. Whilst material assistance is understandably offered to the poor, the media's symbolic power to publicize and confer recognition can also benefit the middle and upper classes in certain situations. At times, television appearances (especially in news rather than entertainment programmes) confer symbolic capital and accrue positive value for the individual. And in moments of exceptional public crisis, television presents itself as a trustworthy and efficient space that becomes a desirable venue for upper- and middle-class charity.

Conclusion: Moralities of Consumption

By focusing on the moment of consumption in the mediation process, this chapter provides empirical material by which we can reflect on media ethics debates on audience ethics and ecological ethics.

This chapter finds that in the Philippine context, the problem of suffering implicates both the witness and the sufferer (see chapter one). The upper and middle classes are implicated in moral decisions to tune in or 'switch off' images of the visceral realities of poverty in the metropolis and ever-present in the mediated centre of network television. Learning about depressing local news and gaining familiarity with *jologs* popular culture contend with aspirational desires that can be tenuously realized within 'zones of safety,' both physical (e.g., gated neighbourhoods, gated universities, malls) and symbolic (e.g., international television, foreign websites on the Internet). Alongside this, people who suffer in poverty themselves seek out local media in their everyday struggles to obtain recognition and redistribution. Whilst Filipino ethnographies discuss how 'those who have nothing' have historically attempted to forge ties with those who have power in everyday life (Cannell 1999; Kerkvliet 1990), in this case the poor finds that clientelistic devotion to television networks offers material rewards.

For both witnesses and sufferers, tension is rooted not in the problem of distance but in the problem of proximity. Whereas Western scholars argue that distance may pose questions of authenticity – whether or not the disaster actually happened and if sufferers truly need aid (Boltanski 1999: 159–61) – proximity in this case cannot but demand some form of acknowledgment. Poverty exists in 'excess' (Tadiar 2004, 92) in and out of the frames of television, and natural disaster always awaits a landmass besieged by typhoons and earthquakes (Bankoff 2003). By coexisting too close to sufferers and disasters, the upper and middle classes are observed to acknowledge suffering, whilst avoiding it through practices of boundary-setting and 'flying over' the suffering that exists around them. By using the term *jologs* as a judgments of taste, they often even skillfully evade moral evaluation by attributing these judgments of value to personal taste. These taste judgments, however, are ultimately revelatory of classed moralities (Sayer 2005). Therefore the branding and re-branding of objects and people as *jologs* must be evaluated as strategic performances by which value is regulated – and hoarded – by elites to maintain status. Because *jologs* is used as a totalizing and pathologizing 'word of hatred' (Skeggs 2005), it is a word that delegitimizes any moral claims of sufferers that may exist within textual representations of suffering. In other words, *jologs* ultimately works to justify the continued existence of the parallel lives of the Filipino rich and poor. Within this discussion of audience ethics then, *jologs* appears as something more than a valuation of other people's cultural capital; it is also a strategic reduction of the moral obligations of the rich to the poor and is suggestive of upper-class fantasy-production of disconnecting from the undesirables around them. The audience orientation of being switched off, whilst indicative of disconnection from the symbolic space of media, is also reflective of social and moral disconnection from sufferers in and out of the frames of television.

Television's hospitality is evident in the ways it provides symbolic and material rewards to the most deserving poor in society. 'Pilgrimages' to the mediated centre (Couldry 2003) are beneficial for sufferers in significant, if not structurally transformative, ways. Sometimes an exchange involves a poor person providing public testimony of suffering in order to reap direct assistance from the media. At other times, it involves poor people being redirected to other authorities because their personal stories sound less exceptional and deserving of recognition and reward. These media practices are likewise subject to moral evaluation by audiences, as lower-class respondents regard this in positive terms, while middle- and upper-classes doubt the sincerity and motives behind them. For some of the middle- and upper-class respondents, these practices give additional justification to 'switch off' local television for 'over-representing' (Wood and Skeggs 2009, 177) and 'exploiting' the poor. However, in some occasions television networks provide an efficient and trustworthy space in which middle- and upper-class charity are mediated in a context where public institutions are regarded – and stereotypically represented (also for the media's own benefit) – as inept and corrupt.

Chapter 4

ENTERTAINMENT: PLAYING WITH PITY

Filipinos also automatically cry, as if on cue, whenever they are interviewed. Be it on the local news [or] some crappy noontime variety show [...] they will flat out whine and cry and tell the whole world that their lives are miserable as fuck and all of their relatives are in their deathbeds and shit.

'Filipino', Uncyclopedia.wikia.com/wiki/Filipino

Having explored audiences' patterns of tuning in and 'switching off' in light of the debates on audience ethics, this chapter pays closer attention to the interpretations that audiences have of television texts that portray suffering. This chapter draws primarily from interviews with different groups of audiences in Manila to reflect on specific debates on audience ethics about the reception of televised suffering. As reviewed in chapter two, audience ethics includes concerns on whether (and which kinds of) audiences express discourses of compassion or pity toward particular cases of mediated suffering. A common assumption here is that emotional expressions of sorrow, indignation or guilt are indicative of a moral concern for the other, but may vary according to class, gender and age (Dalton et al. 2008; Höijer 2004; Kyriakidou 2005, 2008).

Additionally, expressions of compassion toward mediated sufferers are said to be dependent on the production of ideal victim images (Cohen 2001; Höijer 2004; Moeller 1999). This discussion necessarily connects with the debates on textual ethics, where the question of whether to represent sufferers as empowered and humane (Tester 2001) or 'at [their] worst' (Cohen 2001, 183; Orgad 2008, 21) assumes that audience responses are greatly influenced by representational strategies of the media. Media decisions to use close-ups or long shots and to give voice to sufferers or merely speak in their behalf are assumed to have an impact on audience (dis/)engagement with suffering.

This chapter specifically explores audiences' responses to suffering as seen in the genre of noontime entertainment – a local, hybrid genre that mixes the conventions of reality television, game show and talk show. This chapter presents material for reflection as to how a factual genre other than news might have similar or different generic qualities that influence audience discourses of compassion.

At the same time, the discussion of this genre and one popular show in particular – *Wowowee* – was motivated by an ethnographic surprise during fieldwork (Strathern 1999, 9). It was *Wowowee* that respondents routinely pointed out as a television programme 'full of poverty' and 'all about suffering'. It was in this show that the poverty of television takes on its most spectacular forms of over-representation (Wood and Skeggs 2009, 177), with promises also of the resolution of poverty through interventionist and charitable television personalities. *Wowowee* is known for handing out cash prizes, groceries, livelihood packages and even houses and lots to ordinary people who come to the show to play games of chance whilst narrating personal experiences of hardship in confessional-style interviews. By over-representing' (Wood and Skeggs 2009, 177) the poor, *Wowowee* presents a space where poverty is made visible and audible on Philippine television, whilst evoking from audiences contested moral evaluations of the behaviours of its contestants and different lay media moralities of whether television exploits or empowers the poor.

This chapter finds that audience discourses of compassion toward ordinary poor participants in *Wowowee* vary greatly according to class. Respondents across classes express pity or indignation based on different moralities of authenticity, respectability and deservedness. At the same time, they carry different normative expectations toward the media institutions, genres and personalities that mediate suffering. One important finding of this chapter is that normative debates of whether to represent sufferers as having agency or no agency become complicated in everyday life situations, as audiences actually have different interpretations of the agency and deservedness of sufferers on television. However, it also argues that, by taking into account its representational practices and audience responses, we can develop a more holistic and careful ethical evaluation of the ecological ethics of media.

The Filipino Genre of Noontime Show

When advertisers, network executives and media critics categorize *Wowowee* (ABS-CBN; Mondays to Saturdays; twelve o'clock to two o'clock; 2005–2010), they would say that it is a *noontime show*. Whilst classifying a programme based on its timeslot may be unusual in traditional genre theory, noontime show in this case would refer to a Filipino factual entertainment genre that draws from multiple influences.

Wowowee's roots can be traced to the long tradition of vaudeville in Philippine entertainment. Vaudeville, or *bodabil*, encompasses entertainment forms that include singing, dancing, comedy sketches, game show and talent show portions, beauty pageant segments, magic acts and the like. *Bodabil*, in the 1910s to the 1960s, was performed live in small theatres and community centres, but the genre's influence is argued to be evident in contemporary television programmes and live events, such as election roadshows (Tiongson 1994). *Wowowee* is also part of a more recent history of noontime television shows, in which game show segments, comedy sketches and song-and-dance numbers are combined. The game segments are uncomplicated in order to encourage participation from both studio and home audiences. Game segments, hosted by celebrities, are usually variations of name-that-tune, spin-a-wheel, obstacle courses and trivia games, as well as beauty pageants and talent competitions.

Wowowee's unique contribution to the noontime show genre is its overt spotlighting of contestants' personal experiences of suffering. Whereas previous game shows such as *Kwarta o Kahon* (Money or [mystery] box) also featured poor contestants, personal narratives of suffering were not intended to be overtly shared by the contestants. Even if the host's interviews with contestants might reveal their impoverished backgrounds, these were incidental to the main contest.

In contrast, *Wowowee* devotes equal time to the game show mechanics of 'money or mystery box' and emotional interviews with contestants. Live interviews with contestants discuss their families, love lives, work and motivations for taking a chance in queuing up to appear on *Wowowee*. Whilst other game shows engage contestants in conversational repartee, I would argue that *Wowowee* uniquely appropriates the format of the confessional into its game/variety show format.

Over the years, media research has identified the confessional as an intrinsic property and part of the appeal of talk shows. In her discussion of rape narratives on *The Oprah Winfrey Show*, Moorti (1998, 83) argues that 'the act of giving voice to pain' in the form of ordinary people's emotional confessions contains potential for cathartic transformation. She, like Lunt and Stenner (2005), also reflects on the potential of talk shows as 'alternative' and 'emotional' public spheres. Lunt and Stenner suggest that such programmes that hear out 'minority voices' may provide opportunities for moral reflection through the kinds of questioning and judgment encouraged by their format, e.g., 'What have we learned today?' (73). In many ways, this is similar to *Wowowee's* own practices, in which the programme host dispenses advice about proper conduct to contestants with family problems or financial difficulties and addresses studio and home audiences with his personal reflections about right living.

Though some Western scholars acknowledge the confessional for its potential for self-reflection and democratic speech, in the Philippines, it is in fact at the crux of the vitriol against *Wowowee*, particularly from academics and cultural critics. They bemoan *Wowowee*'s tendency to 'glorify poverty' and 'wash dirty linen in public' by encouraging the poor to share their stories of suffering: 'People are conditioned to being poor. People are made to think that it's good to be poor,' according to Koh (2006). *Wowowee* is after all known for having themed episodes where the day's contestants might be all housemaids, or disabled people, or street cleaners or recently laid-off overseas Filipino workers. Special groups of poor people are invited to describe their lives and the difficulties that they have experienced, and – based on the pity they evoke from their stories – may be given 500 pesos (7 pounds) on the spot even before they participate in the actual game show segments. Some contestants are even asked to demonstrate talent for cash, so the sight of paraplegics doing acrobatic tricks or toothless old ladies doing comedic interpretations of popular songs is not uncommon. By narrating and spectacularly performing their suffering on television in order to win cash prizes or garner fame, these contestants are assumed by critics to be 'stripp[ed] of [...] their last ounce of dignity' (Dancel 2010). Once again, television's practice of over-representation (Wood and Skeggs 2009, 177) is read here as actually being abusive or patronizing rather than sincere and beneficial.

Critics recognize that the success of this 'creative' format is rooted in lower-class television audiences' identification with the poor people onscreen. *Wowowee* supposedly offers 'false promises of salvation' (Fonbuena 2006) in perpetuating the expectation of dole-outs and the myth of *biglang yaman* (sudden fortune) to television audiences likely to be in conditions of poverty themselves. Certainly, the programme is unlike other reality/game shows on Philippine television that may:

> require actual talent, intelligence, or purchase of sponsors' products [...] [in *Wowowee*] all one has to do [...] was to fall in line outside ABS-CBN. If host Willie Revillame spots you, then you are in luck [...] If [he] fails to spot you, at the very least you get fed with biscuits from sponsors, which is not at all bad considering you have nothing at home (in Get Real Philippines 2011).

Shows such as *Wowowee* illustrate the consequences of a profit-driven media environment that caters 'to that audience at the bottom of the pyramid [...] target[ing] the least common denominator. It is like feeding one kind of food, the easiest to digest, yes, baby food, to adults with teeth' (De Jesus 2011). Cultural elites' infantilized image of the television audience extends to their perception of the *Wowowee* studio audience participants who win money simply 'by being there, for doing nothing' (ibid.), with minimal reflection about the physical and emotional labour that media pilgrims may endure.

Ultimately, critics interpret *Wowowee*'s apparent generosity as a marketing strategy to keep its followers indebted to the network and thus remain loyal viewers for ABS-CBN's ratings success (Doyo 2006). To many critics' dismay, *Wowowee* consistently garners high television ratings, driven by the low-income viewers that comprise the majority of the viewing audience. *Wowowee* was recorded during fieldwork as having the second-highest ranking in nationwide ratings for daytime television programmes, behind only one other ABS-CBN game show, *Kapamilya Deal or No Deal* (McCann Erickson 2009).[1]

Pilgrimage and Ritual: Game Show in Public Life

During one of my fieldwork-related trips to ABS-CBN, after an early morning interview with an executive, I walked past the waiting area for *Wowowee*. It was around eleven o'clock, and by then there were about a hundred people in line. They were in a cordoned area, on the sidewalk of a busy one-way road; a tarpaulin provided shade to the enterprising early risers, but latecomers were left to bake under the midday sun. People were chatting in small groups, women fanning children from the 35-degree summer heat. Whilst their faces betrayed fatigue, perhaps from the long wait or half-day journeys from faraway provinces, they were also alert, excited. Necks would crane over the steel railings when a Mercedes Benz would drive by or when a tall man wearing sunglasses would cross the street. Who knows which celebrity they might spot? A middle-aged lady was leaning on the railings stood near the pavement where I was walking; I was hesitant at first, but when she caught my eye I stopped and started a chat. I asked why she was in line and she replied, 'Oy! You can never tell! It might just be my lucky day!' Apparently, that week, no one had won more than 50,000 pesos (700 pounds) in one of the segments, and the odds of winning big that day were pretty good.

Such thoughtful calculations and hopeful sentiments could have been the same as those of the 30,000 *Wowowee* devotees who had gathered in Ultra Coliseum in February 2006 (Torres 2006). For its first anniversary episode on 4 February 2006, *Wowowee* promised to give away millions of pesos, a house and lot, a car, taxis and passenger jeepneys[2] to contestants to thank them for the show's ratings success. Rumoured promises of 20,000 pesos (300 pounds) to the first 500 people in line drove tens of thousands of people to queue and try their luck (Coronel 2006b). Local television has a long history of game shows offering cash and gifts, but none as lavish as *Wowowee*. Whereas other shows would raffle off livelihood packages, such as variety store starter kits or passenger tricycles, the stakes in *Wowowee* were exponentially more astonishing, more extravagant and more instant.

Days before the event, people flocked to the stadium, bringing with them their entire families, carrying only small pillows to sleep on, fans to cool down from the heat, and not much else. Some came from the islands of Visayas and had only enough money for a one-way fare – that was how confident and hopeful these devotees were, journalists claimed (Cruz 2006; De Quiros 2006). It was reported that some people snacked only on peanuts, fish balls and candy, whilst some, as if on a spiritual journey, ate nothing and drank only water (Inquirer.net 2006). It was also reported that there were no toilets; afraid to lose their places in the queue, people would leave briefly to urinate and defecate by the sidewalk on the other side of the street.

Reports say that at two in the morning, ten hours before the show, television producers announced over the megaphone that they would hand out tickets to only 17,000 people – the capacity of the stadium. As a consolation prize for those who would be unable to enter, they would hold a random drawing to to givee cash to the people in line. The announcement got the crowd of 30,000 agitated, as people started pushing, shouting and squeezing their bodies through the slim openings of the steel gates, desperate to cross over to safe ground. Finally, at six o'clock, six hours before the show, a throng jostled and pushed, causing a melee, forcing the gate to crash and starting the stampede that would eventually kill 71 people and injure over 800 others (Coronel 2006b).

I remember waking up to the news in disbelief. I remember watching on television a scene 'straight from a war, a killer earthquake, or a landslide in a far-flung corner of the country', in the words of the columnist De Quiros (2006). By then, the shots were of mothers crying, bodies of the dead piled on top of each other by the collapsed gate, shoes, slippers and children's toys littered all over the street. I remember ABS-CBN reporting on the tragedy that befell their show and its viewers. I remember GMA reporting cautiously and calmly, so as not to appear biased, indignant or gloating to its network rival. And I remember being astonished when, at around noon, after a long commercial break, the scene suddenly changed over on channel two. We were outdoors no more. The news ticker and the split screen set-up between news anchor and field reporter were gone. We were now inside the stadium – a very packed stadium. *Wowowee* was on. Though the set was empty of its cash boxes, colourful roulettes and bikini-clad dancers, the show's host Willie Revillame was present. He was out onstage. He was in tears. He spoke about the tragedy and offered a prayer. He only wanted to make people happy, I remember him saying, as he often does.[3] He explained, to the disappointment of a still-eager crowd, that unfortunately, the show would *not* go on that day. After half an hour of tearful speech from the host, the show ended, cut to a commercial, and the afternoon soap opera came on.

Over the next few days, ABS-CBN and *Wowowee* came under much scrutiny [...] or at least as much publicized scrutiny as could be made.[4] Official investigations were carried out by the government and police. Then interior undersecretary Marius Corpus was quoted as saying, 'offering so few tickets to so many people can be likened to throwing a small slice of meat to a pack of hungry wolves' (Torres 2006). ABS-CBN set up a foundation called 71 Dreams to provide cash and burial assistance, as well as psychological and spiritual support, to victims' families. All but three of the 71 killed were women, almost all elderly. And *Wowowee*, its meanings, relevance and overall reason for being, came under attack.

The official editorial of top newspaper *Philippine Daily Inquirer* blasted how *Wowowee* 'was exploiting the poor to build higher ratings' by promoting 'games of chance' as 'solutions to poverty' instead of 'hard work' (Inquirer.net 2006). Coronel (2006a), of the Philippine Center for Investigative Journalism (PCIJ) attacked host Revillame for acting the role of 'messiah of the idiot box'.

Middle- and upper-class bloggers, referencing Marx, lamented how *Wowowee* had become the opium of the Filipino masses. As a whole, critics and academics understood – and represented – *Wowowee's* followers as victims. They are victims in that they possess little agency: maltreated as pilgrims and refugees in the mediated centre, shamed as sex objects, made fun of as contestants, used for other people's entertainment and business motives and – in the wake of the tragedy – treated, even slaughtered, like animals as a function of their poverty.

This debate was hardly consensual, however. *Wowowee's* many devotees, most from the lower class, defended Willie Revillame and ABS-CBN (Sir Martin 2006). *Do not blame Willie*, they said. *It was merely an accident. Who knows God's plan?* As eventually reported, to little surprise, only 16 people would file complaints to the National Bureau of Investigation (NBI) (Santos 2007), and no one from ABS-CBN was jailed or fired.

Public criticism was also directed towards the poor themselves. ABS-CBN's head of security pinned the blame on the 'unruly crowd' (The Freeman 2006). A GMA journalist reflected on whether anarchy is 'ingrained in our culture' because of Filipinos' seeming inability to queue properly and behave in a crowd (Severino 2006). Many in online fora and blogs blasted the attendees' 'mob mentality', their 'greed for "easy money"' (Jimenez-David 2006), and their 'shamelessness' (*garapal*) in insisting that that the show should continue in spite of the pile of dead bodies outside the stadium.[5] Others lamented the poor's fatalism and lack of resourcefulness, shown in their overdependence on 'lotteries, raffles, txt to win, etc. to make their lives better [...] [They] don't think of sending their children to school anymore for them to have a better and secured [sic] future' (in Coronel 2006c).

As succeeding sections illuminate, this chapter finds that people's responses to *Wowowee* are revelatory of moral discourses both about suffering and the processes by which television mediates suffering. Moral discourses of authenticity, deservedness and respectability reveal not only classed judgments of right conduct and personhood but are likewise shaped by the mediation process. Specifically, practices of queuing up at the 'mediated centre' (Couldry 2003) and performances of the self on live television invite different kinds of moral judgments based on class backgrounds and interests.

Cheers of Compassion

During fieldwork, *Wowowee* was a frequent talking point among respondents of all classes. When I would ask upper-class respondents how and where they see suffering or poverty, *Wowowee* was the most frequent response, followed only by mentions of street children who knock on their car windows begging for alms and their own live-in housemaids or nannies. Middle-class respondents would meanwhile cite *Wowowee* before the news when I would ask for examples of television programmes they know that portray suffering. And whenever I visited the Park 7 community around noontime, I would frequently hear the unmistakable songs and sounds of the show echoed by the many television sets all tuned in to the same programme. Unlike soap operas and other factual entertainment shows on local television, *Wowowee* is unique in that audiences across all classes have at least heard about it and can identify its celebrity hosts. Whilst *Wowowee* has a recall value that other local programmes do not, upper-class respondents usually speak about it in very critical terms. Many of them mimic critics' and academics' evaluations of exploitation in how a television network uses the poor to surge ahead in the ratings war. Middle-class respondents cite the values of fatalism and false hopes that the show propagates. Both upper- and middle-class respondents cite the *Wowowee* stampede (two years in the past at the beginning of fieldwork) as a significant reason why the show has negative value. The fact that no ABS-CBN official and no *Wowowee* staff member was held legally accountable by the courts is common knowledge to upper-class respondents and informs their overall negative evaluation of the programme. Additionally, they make comparisons to American television programming that they are not only more intimately familiar with but also cite as having more value than this local show.

NATASHA: I can't believe that *Wowowee* can go on like this. Like, do you think *American Idol* could continue on after a horrific stampede? [The Americans] would surely axe it – and the officials behind it!

(Group interview – private university)

JONATHAN: What do you think makes *Wowowee* so popular?

KATRINA: Maybe it's also like *Jerry Springer*, you know? The fact that it's just so over the top? I don't know really. But I think people watch

it for the [uses fingers to do air quotes] *charity*. So maybe it's
even more similar to *Oprah* then. Oh God, no offense to Oprah.
Oprah will likely have a heart attack if she watches *Wowowee*.

(Group interview – private university)

However, some upper- and middle-class respondents admit to being
entertained by one particular segment of this two-hour long show. They
particularly cite one game segment, 'Hep Hep Hooray', as funny and enjoyable
because of its old-school parlour game mechanics.

'Hep Hep Hooray' is a contest featuring ten participants selected from the
live studio audience of five thousand. This selection process is carried out
in the segment before the actual 'Hep Hep Hooray' segment; this process
involves the live studio audience dancing, even gyrating, in front of their
seats whilst the camera pans across the studio and selects ten participants.[6]
'Hep Hep Hooray' officially begins with the ten participants lined up in one
neat row facing the main camera. They must alternately and quickly exclaim
'Hep Hep' or 'Hooray' as the host randomly points the microphone to the
contestants. Additionally, exclaiming 'Hep Hep' or 'Hooray' comes with
appropriate gestures: contestants must clap their hands against their knees
on 'Hep Hep' and raise their arms in the air on 'Hooray'. The mechanics are
very simple but done very quickly; within 10 minutes, the cohort is quickly
whittled down. Eliminated participants receive a gift pack containing soap,
detergent and snacks provided by the show's corporate sponsors. The one
winner of 'Hep Hep Hooray' moves on to a longer segment, in which this
winner is made to choose between cash or a sealed mystery box, with as much
as 500,000 pesos (700 pounds) up for grabs.

During my group interviews with upper- and middle-class respondents,
'Hep Hep Hooray' is the one rare segment that elicits positive comments.
Vanny, a magazine editor, admits to enjoying the 'children's party' aspect of
the show but nevertheless distinguishes this enjoyment from that of *Wowowee*'s
more rabid fans.

Indeed, unlike other segments of the show where participants dance and
gyrate for the chance to win money or where they are interviewed about
personal stories of hardship, 'Hep Hep Hooray' emphasizes participants'
physical skills of speed and coordination. It is also perceived by my upper-
(and middle-) class respondents as a more or less level playing field, where
any of the ten has the chance to win and where winning is based on the
personal abilities of the contestant rather than on the subjective judgment
of the programme host. Certainly, the mechanics of this contest resonates
with their genre expectations that a game show should provide fair and equal
opportunities for contestants to succeed.

In contrast, lower-class respondents received this segment very differently. Lower-class respondents were much more actively engaged in the process by which the ten contestants are whittled down to one finalist. Lower-class respondents cheered for particular contestants in the line-up of ten. When I asked my lower-class respondents to retell the segment, they expressed concern for and evaluations of individual contestants.

JONATHAN: What can you say about what you just saw?

ESTHER: Too bad! I was really hoping the old lady would win.

JONATHAN: Oh, why do you say that?

ESTHER: I can see it in her eyes. The pain in her eyes. She needs money. She needs help.

JONATHAN: How can you tell who needs help?

ESTHER: It's in the eyes. In the face [...]

JONATHAN: So you think the winner is actually doing okay? How?

ESTHER: Yes. From how she carries herself. How she walks.

(Group interview – park 7)

JONATHAN: How did you feel about the segment?

MARGIE: It was okay [...] But I was hoping the other contestant would win!

JONATHAN: Why?

MARGIE: Because, you know, she was more pitiful.

JONATHAN: Because?

MARGIE: Nothing. Um, you can see from the colour of her skin that she wasn't well-off, probably a laundry woman who washes her boss' clothes under the hot sun. The girl who won looks like she's okay financially. She speaks good English [...] She's pretty.

(Group interview – Mandaluyong)

Lower-class viewers place great significance in the selection of who gets to play and who gets a chance for money. Whilst they recognize that to be selected is partly a function of luck – of being at the right place at the right time ('You never know!' as the woman lining up for *Wowowee* had told me) – selection also has a deeper significance. To be selected means not only being able to compete for prizes, but it also means being seen and heard over other people who also experience great hardship in their lives.

For lower-class viewers, there is high emotional engagement as to who wins and who loses. They are fully aware of the stakes of the game: to be selected means having a chance at instant material gain that the media are able – and expected – to dispense more than any other institution in Philippine public life.

For upper-class viewers, there is little interest or consequence as to who wins in the segment:

JONATHAN: Were you rooting for anyone?

NATASHA: No.

JONATHAN: But didn't the winner look a little more well-off than the other?

NATASHA: Um. The way she did her makeup shows that she's poor. The way her skin looks, you know? Yes, the grandmother looked poor too. So they're both poor.

(Group interview – private university)

For upper-class respondents, the consensus is that people in *Wowowee* are all poor, and therefore equally deserving of any of the show's prizes. In this excerpt, a private university student with limited exposure to suffering in her life and limited consumption of Philippine media is revealed to totalize the *Wowowee* studio audience by perceiving them as equally, rather than variably, poor. Unlike lower- and middle-(and some older upper-) class people who have had direct and mediated experiences with poverty,[7] most upper-class audiences are less likely to draw from a deep pool of indicators of suffering to resolve whether one contestant may be more poor and/or deserving than the other. Additionally, their negative evaluation of the programme as exploitative influences their lack of concern for the fates of individual contestants: the contestants' willing participation in an exploitative and degrading show taints their evaluation of the moral value of contestants themselves. This prevents them from articulating any 'discourse of compassion' (Höijer 2004), as suffering is largely retold in homogenized terms, or reduced to a manipulative 'textual game' (Chouliaraki 2010, 120).

Middle-class responses are much more complex, as my respondents can often relate to experiences of suffering. They or someone close to them have undergone difficult life experiences and are knowledgeable about poverty. As mentioned in chapter three, the middle-class in the Philippines occupies a dangerous and slippery position (Parreñas 2001; Pingol 2001), in which people can easily lose this status as a result of personal tragedy or external events. Middle-class respondents, whilst sometimes claiming to feel pity for some contestants whose experiences resonate with their own, also at times assert moralities of resourcefulness and hard work that they consider necessary to climb out of poverty – observed too by Pinches (1999) in the context of the Filipino 'new rich' contesting 'old rich' cultural dominance. Though not all of my middle-class respondents believe that *Wowowee* is a degrading and exploitative programme, they challenge its contestants' overdependence on games of chance as solution to poverty. They argue that contestants' pilgrimages to *Wowowee* should not be a substitute for real labour, citing their own life stories as examples of personal transformation achieved through hard work. As Jessica, a 33-year-old small shop owner, said,

> I don't think it's absolutely wrong to gamble, or buy a lottery ticket, just like there's nothing completely wrong with lining up for a show. But even in the small chance that you win, it's likely that you won't know what to do with the money because you didn't learn anything along the way. That's the difference with hard-earned victory.

In contrast, lower-class respondents articulate discourses of compassion, particularly 'tender-hearted compassion' – the expression of sorrow and empathy for victims (Höijer 2004, 522). Tender-hearted discourses of compassion are directed to the specific contestants of *Wowowee* that they evaluate as poorest and most deserving. These discourses are accorded individually rather than collectively, as they take active interest in evaluating individual contestants' physical appearances and (later) listening attentively to their stories of hardship. The lower-class respondents are more sensitive to contestants' identities and varying conditions of suffering, and they respond emotionally to people that they find touching. In a way, there is this deep sense of *damay* (mourning; sympathizing) where a contestant's stories are reflected upon as their own, which is very different from my upper-class respondents' collectivized appraisal of contestants ('They're all poor'/'They all need money').

In the context of the 'Hep Hep Hooray' segment, where ten people are lined up in one neat row facing the television screen, lower-class (and some middle-class) audiences' discourses of compassion depend on a process of reading indications of poverty and suffering from contestants' bodies. Popular stereotypes about skin colour, body frame, age, clothing, accent and race/ ethnicity inform judgments of participants' physical appearances and whether they are sufferers or not.[8]

My interviews also unearthed subjective indicators that were less overtly evident than physical or material indicators. Some lower- and middle-class audiences enumerated how they could tell whether people were poor based on 'their eyes', 'the way they stand' and 'how they act in front of the camera'.

My respondents claim that a life of hardship and poverty could actually be read off a person's eyes and whether they are sad, wrinkled or worried. Whether people stand up straight or slouch sheepishly, especially when interviewed by the programme hosts, is also assumed to convey the presence or absence of confidence and self-assuredness that people possess in different degrees. Certainly, well-off people are assumed to act with more confidence, especially in the generic space afforded by the game show in which they have little concern for the success or failure of their actions.

Nevertheless the distinctions that lower-class audiences make between poor and not-poor, deserving and undeserving contestants are ultimately hinged on subjective evaluations. Thus, through the course of watching *Wowowee*, audiences sometimes discover that contestants that they considered poor/deserving based on their physical appearance were not actually poor/deserving after listening to their interviews in the later segments.

I would argue that it is precisely the unpredictability and uncertainty of the authentic poverty and deservedness of *Wowowee* contestants that makes the viewing experience exciting and entertaining. The confirmation of a hunch about the contestant's suffering whether correct (*Sabi ko na nga ba!* 'I knew it!') or incorrect (*Di ko akalain*. 'I wouldn't have guessed') brings fans the most excitement in viewing. Because authenticity and deservedness of sufferers are in constant play due to programme and genre conventions, it is the unpredictable affirmation and contradiction of the moral judgments of *Wowowee* and its audiences where narrative climax – and audience enjoyment – ultimately lie.

Authenticity and Deservedness

Audiences work through judgments of authenticity and deservedness not just in the game show-driven segment of 'Hep Hep Hooray', but also in the succeeding confessional-driven segments. After one contestant wins 'Hep Hep Hooray', s/he moves on to the 'Cash? *Bukas*?' (*Cash? Open [Mystery Box]?*) segment. Here the game mechanics entail the finalist to first select one mystery box from 10 boxes. One box contains up to 500,000 pesos (7,000 pounds), two others contain less than 100,000 pesos (1400 pounds), one contains a 'livelihood package', and six others have dud items such as toothbrushes, potatoes or rubber slippers. At the same time, the programme host tempts contestants to take home cash amounts instead of the mystery box. The cash amounts dangled in front of contestants begin at 10,000 pesos (140 pounds) and end at around 50,000 pesos (700 pounds).[9] In between these choices of 'cash' or '*bukas*' (open), the host engages the finalist in an emotional interview about his/her life. Consider this exchange in the middle of the segment:

Wowowee (Episode: 2 June 2009)

WILLIE: What will you do with 300,000 (pesos) in case you win?

GIRL: We'll use it to pay for my brother's tuition fees in school. [tears up]

W: Mmm […] She's crying now [….] She's crying.

[audience giggles audibly]

W: So why are you crying?

G: [shakes head, covers face]

W: Where's your brother?

G: He's at home now. [suppressing tears]

W: Why, why are you crying […] What's your mom's work?

G: She's in Dubai.

W: What is she there?

G: In the tiles (industry).

W: And your dad?

G: Tricycle driver. [wipes tears]

W: [turns to audience/camera] Gosh, my heart just fell. [Girl gives friendly nudge to Willie]

W: [To the audience] The dad is a tricycle driver, the mom is working in Dubai. [Turns to girl] Are you not in school now?

G: No, I stopped so I could work for them. [wipes tears]

W: [To the audience] Such a good girl. [Then looks to girl with sadness/pity] [audience applause -cued]

[cut to audience applauding]

W: [rests arms on mystery box] Sigh. [gives wail, feigning tears, then breaks to laughter] Oh such is life! [wipes eyes] Okay, good luck to you! [To *Wowowee* bikini girl holding the money:] Come here! [To contestant:] Congratulations. Hope this will turn out good […]

[they play a few rounds of 'Cash? *Bukas*?']

W: So what do you want to say to your mother? [turns to contestant] [sentimental music plays]

G: Mmm, Mom, I hope you are watching… Don't overexert yourself. I hope you're thinking of us there […] I hope you're thinking about our situation… [wipes tears] Always take care of yourself there. That's it. [Turns to Willie]

W: What about for your dad? In the tricycle now?

G: Umm, Papa, I hope you take good care of the tricycle I bought for you. [Turns to Willie:] I was the one (who bought the tricycle for him). I had to do it because he does nothing but sleep all day. [audience laughter]

W: Oh. So tell your dad! [Turns to the camera] You are so lucky that your daughter was able to purchase a tricycle for you. [turns to the girl] Where did you get the money for the tricycle? Where will you be able to get the money? When you're not studying? What's the truth?

G: I worked for it. [purses lips]

Such an exchange is typical on *Wowowee*, where contestants are seen by viewers as '*mababaw ang luha*' (prone to tears or even ready to cry, at least according to the Unyclopedia.com entry on 'Filipino' quoted at the beginning of the chapter). My respondents, across classes, view this display of emotions on *Wowowee* as essential for contestants if they wished to gain money, or extend their airtime on the show. Having a good story, and being a good narrator of one's hardship, are important skills to gain the sympathy of the programme host, who may immediately dispense money to a contestant whose personal story moved him emotionally. In certain occasions, after a contestant's emotional interview, Willie Revillame may even guide contestants to make correct choices to win bigger prizes. In 'Cash? *Bukas*?' for example, Willie can drop hints as to which box to pick, or persuade contestants to switch their choice of boxes, or insist that they take home the cash instead of the box. His benevolent intervention is not equally available, however. Some contestants find themselves left to decide their own fate.[10]

Wowowee's mechanisms of suspense typically revolve around ambition and risk-taking: whether one should remain content with the certainty of the present yet modest reward, or whether one should 'risk it all', 'jump at the chance' (*makipagsapalaran*), 'leave it all to God' (*bahala na ang Diyos*), and select the mystery item. *Wowowee's* host is intimately involved in this decision making, as he himself plays judge and ascertains the contestant's need relative to the prize at hand. The decision to take cash or mystery box is framed not simply within the game show context of winner/loser but, through the device of the confessional, is simultaneously incorporated within real-life conditions of hardship. In the example above, a twenty something, fair-skinned female reaches the 'Cash? *Bukas?*' round after besting nine other contestants in 'Hep Hep Hooray'. Here she fields Willie's opening questions, which typically revolve around the purpose in which the cash prize might be used for. She is asked about her family situation, and the themes here – education, occupation, separation from loved ones – are typical in the show.

Interviews are emotional and involve tears. They are stories of loss, disappointment, danger and death: unpaid hospital bills, unplanned pregnancies, job lay-offs, housing demolitions, bereavement, oppression by employers both here and abroad. Willie, or *Kuya* Willie (Big Brother Willie, as he is called by many contestants), positions himself as an older, richer sibling or uncle concerned with contestants' lives. He offers consolation, encouragement and judgments of their conduct, along with material rewards. For instance, when the contestant in this excerpt came to reveal that she worked as a 'dancer in a club'[11] to help sustain her family, Willie offered support, used humour to lighten the mood, and granted her 'absolution' from shame and harsh judgment of audiences.

[continued from previous excerpt on page 103]

W: No, I understand that. [places arm around girl's shoulder] [Turns to the audience] That's okay. That's the way life is. [Turns to girl] Where is that place?

[girl laughs]

[canned laughter plays]

W: I haven't been there. I haven't been there. Is *that* expensive there? [laughs] [girl gives playful shove]

W: No, that's okay. [Turns to the audience] This type of person we should not condemn. [puts arms around girl] We should admire her because [twinkling sound effect] […] She works in a (strip) club

– that's okay! It's nothing. So she can sustain her family. We should not think of this as bad. [cut to audience applause – cued]

W: That's okay. [Turns to girl] As long as you know that's just a job. And then after your work from the club, all you're thinking about is bringing home money for your parents […] [close-up of girl wiping tears]

W: […] So you can pay for your brother's education. We know that. Many are in the same situation as you are. People like you are not to be condemned. The more we should try to understand and even love [turns to audience] because this type of people are lacking in love. Let's give a round of applause for those like Danica. [musical cue] *20,000 pesos!!! Danica! Cash? Bukas?*

The language is therapeutic and moralistic. It is about rights and wrongs, and is laden with implicit judgments about the right and wrong ways to cope with suffering. In this example, the contestant receives judgment whereby her profession as a prostitute is divorced from her intrinsic moral worth ('As long as you know that's just a job'). Her choice of profession is interpreted and justified by Willie as merely instrumental in achieving a higher value of being a dutiful daughter and providing for one's family *through whatever means*.

Viewer responses

In viewing this segment, some upper-class respondents denounced the desperation of 'the poor'. Some fixated on the *laziness* of the girl's father (whom the contestant mentioned as habitually sleeping all day). This would resonate with Höijer's (2004, 522–3) 'blame-filled compassion' category, which identifies possible causes or perpetrators of the suffering of others.

Lower-class respondents' engagement with her story was much more complex. They narrated at first their stance of skepticism towards the authenticity and deservedness of the contestant. They had initially judged her as an undeserving contestant based on her physical appearance of being fair-skinned, young and attractive. They claimed that she did not look more deserving than the dark-skinned older woman whom she beat in 'Hep Hep Hooray'. But from the contestant Danica's storytelling, lower-class audiences became convinced that they had made a mistake in their initial judgment of her. Because of her storytelling performance, they reevaluated her as actually *truly deserving* of the prize. They express tender-hearted compassion through articulating a sincere hope for her to succeed (*Dapat siya manalo.* 'I hope she wins').

In this instance, most middle-class respondents had similar responses to lower-class respondents, as they themselves identified with the contestant's personal struggle to help her family out of poverty. Danica's story and storytelling abilities enabled identification, and they expressed tender-hearted compassion for her. However, those middle-class respondents with a prior general critical evaluation of the programme for its (perceived) de-valuation of real work ethic remained skeptical of how Danica's and her family's life would in fact be helped by the programme's 'generosity'. Older middle-class respondents cited, for instance, how inflation has reduced the value of money, and whilst *Wowowee*'s cash prizes sound like large sums of money, they surmised that Danica's winnings would be quickly used up in the payment of debts and requests for assistance from other family members and neighbours. Younger middle-class respondents were less cynical and more likely to express interest in Danica's fate in the show.

In this segment then, lower- and some middle-class audiences' judgments of suffering move from the level of appearance to the level of speech and narrative. Whereas in the previous segment viewers could merely 'read' suffering from contestants' bodies, here additional criteria come to play in some audiences' continuing attempt to scrutinize and identify who is the deserving sufferer.

Perhaps it is best to pause here and reflect on the local meaning of 'deserving' in the Tagalog word *nararapat*. *Nararapat* finds its roots in the word *dapat*, which means 'should', related to the word *sapat*, which means 'appropriate' or 'enough', and also informed by the value of *tapat*, which means 'honest' and 'trustworthy'. In idiomatic expressions of *nararapat* (*Dapat siya manalo / Tama na siya ang mabigyan / Totoo siya*) ('S/he is deserving to win'/'It is rightful s/he is given money' /'S/he is true and trustworthy'), we find a telling clue that, in people's engagement with suffering around them (whether mediated or face-to-face), the question of deservedness as fairness is always-already linked with questions of deservedness as authenticity, sincerity and trustworthiness. This is why, in people's engagement with *Wowowee*, of central concern for audiences is whether and how much sufferers are *plastic* (fake) and *parang artista* (like actors). People tend to be skeptical at, for instance, contestants who are too quick to cry or too open as storytellers, as if speaking from a prepared script. The contestant above, for instance, in her coy and tentative manner of talking about her job and situation was successful in gaining my respondents' trust.[12] 'If you are too dramatic and show yourself as too pitiful, then it's probably just an act. You have to show that you have fight, that you're a survivor, so that you can be trusted,' Eloisa, a housemaid, shared with me. Her friend Lita shared, 'If [your story] is too dramatic, then it becomes not only doubtful [...] It's also likely fake. If I'm suffering, then I have to do something. That's what Willie wants to see.' Some lower-class males were in fact critical of displays of grand emotion and tears by programme contestants – especially of male finalists. Male finalists are judged too on standards of masculinity, including the ability to manage one's emotions in public, and are expected to narrate his record of being a good and loyal provider for his family. Viewers' normative standards of 'deserving male sufferer' on *Wowowee* is in many ways consistent with meanings of (and social pressures to be) a good father or husband (Pingol 2001).

Additionally, this question of authenticity is confounded by the programme and genre conventions at play. Hill (2005) has previously argued that authenticity is a quality sought for by audiences of reality television, where producers' claims to reality contend with the motivations of reality show participants as well as perceived producer manipulations to create suspense and drama. In

Wowowee however, I find that lower-class audiences tend to doubt more the participants' sincerity rather than the producers' manipulations. There is a belief that because of poverty, (some) poor people become desperate to get ahead of poorer and more deserving others and may go so far as manipulating rich patrons (like the media) for assistance. For upper-class respondents though, the question of authenticity in a hybrid reality show like *Wowowee* does not come into play, given their assumption that everyone is equally bankrupt, economically, culturally and morally within this mediated space.

So far then, audience interpretations of two segments of *Wowowee* have revealed to us how (some) audiences may work through judgments of the deserving sufferer. People make judgments of the deserving sufferer using two intimately linked criteria: the criterion of truth and the criterion of fairness. Within the criterion of truth are visual and auditory indicators such as skin colour, body frame, age, clothing, accent and ethnicity that should, taken altogether, hint at authentic poverty. And within the criterion of fairness lie moral, culturally and religiously informed judgments about the gravity of suffering that the person has endured, the amount of work that she/he has done, the resourcefulness that she/he has displayed to cope with the suffering and the possible consequences that winning might bring to her/his life.

As Jocano (1997) and McKay (2009) have noted, the expectations to fulfill obligations to others, especially to kin, is used as a central measure by which Filipinos judge the moral worth of individuals.

Agency and Strategic Suffering

In this section, I argue that judgments of the deserving sufferer are underpinned by understandings of agency and victimhood. As previously mentioned, media ethics debates in textual ethics variably argue whether sufferers on television should be depicted as having agency or less agency, with the assumption that representational strategies influence audience engagement towards televised suffering. Drawing from my interviews though, this section argues that agency is not solely a property of the television text, but is dependent on audiences' interpretations. Audiences, from different classes, differently perceive the presence or absence of agency in *Wowowee* 'sufferers'.

Let us begin with upper-class audiences. From the earlier discussion on 'Hep Hep Hooray', we learned that upper-class respondents are not concerned with the success or failure of one contestant over another. I suggested that this disinterest comes from an understanding of the show's contestants as homogeneously poor and therefore equally deserving, rather than variably poor and variably deserving, as lower- and middle-class respondents understand. Certainly, they draw from limited resources in recognizing and understanding differences among sufferers, as they inhabit zones of safety and 'fly over' zones of danger, in the desire to construct borders by which class differences are blocked or at least managed. In chapter three, we saw that boundary construction in symbolic spaces of the media follow geographical boundaries between rich and poor in urban life.

Upper-class respondents make frequent references to *Wowowee*'s exploitation of contestants and manipulation of game show mechanics that run contrary to their genre expectations.

Perhaps this is most evident in their complaint about *Wowowee*'s sound production. A common complaint I heard among my upper-class respondents is how the show is 'so noisy'. Some viewers find Willie too talkative: 'He lectures contestants how to live their lives. Umm, does Ryan Seacrest do that?' Some dislike the sound of his laughter: 'When he makes fun of the contestants, his laughter is so irritating. It's demeaning.' This surveillance of excessive emotional displays is by no means local to the Philippine context; Bourdieu (1986) and Skeggs (1997; 2004) have made similar observations in the French and British contexts.

Further, upper-class respondents find the sound crew of *Wowowee* intrusive, insofar as their sound effects and background music call attention to themselves. In one group interview, the sound production became a heated point of discussion.

> HAPPY: Why do they need to play melodramatic music when a contestant starts crying?
>
> DEBBIE: I know right. This isn't a telenovela. It's as if they're cuing them what to feel.
>
> CARLO: And the contestant too [is being cued] [...] This is so fake.
>
> DEBBIE: At least in a drama, you know it's all a show. Here you don't know right from wrong.
>
> (Group interview – private university)

In the same episode, Willie Revillame was shown losing his temper with the show's sound crew. 'Hey! Please take this seriously. You guys are so slow! You're not listening!' he calls out to them, half-serious in his tantrum when the crew apparently failed to play music at the beginning of a confessional interview.

> HAPPY: So he's angry they forgot to play the drama music?
>
> CARLO: Well, he can also get angry when they forget to play the [canned] laughter
>
> [sound effect].

HAPPY: To me, it's insensitive to play that music. And to get angry in front of everyone too! It's, like, the contestant is being used for show.

JONATHAN: What do you mean? She's made fun of?

HAPPY: Yes, in a way. Because she's in the middle of telling her story. And by playing music, you feel like her story is something from a drama.

JONATHAN: You think the emotions are artificial?

HAPPY: Yes, it's like, 'my story is sad' and then the show manipulates you to feel pity for her [...] When that should be natural. It's supposed to be natural, truthful.

(Group interview -private university)

What is objectionable for them here is how the sound production seemingly manipulates contestants and audiences to cry, to feel and to release emotions. When melodramatic music is played during an interview, they find the contestant 'disrespected'. Their expectation of a game show is for the show to focus on game elements and to provide contestants opportunities to win prizes in games that test personal skill (such as in their preferred segment of 'Hep Hep Hooray'), rather than through contestants' ability to evoke pity. For my upper-class participants, the contestants are used as the actual objects of entertainment in a programme only interested in television ratings and advertising revenue. Whilst *Wowowee's* sound production here is insistently encouraging of the audience's identification, upper-class audiences resist this demand. They find that the show's obtrusive sound production not only violates reality/game show conventions of arriving at 'really real' moments (Couldry 2003, 125), but also trivializes people's real conditions of suffering by using fictional entertainment genre conventions, such as melodramatic music.

All in all, these perceptions (of the contestants as all generally poor, of the media space as a *'jologs* space' and of the programme's genre-bending and overt manipulation of emotions) reveal an implicit judgment of *Wowowee* participants as *victims*. As victims, *Wowowee* sufferers are understood to possess little agency. They are indistinct, all belonging to the abstract aggregate of *masa*, all worthy of equal amounts of compassion, if minimal actual attention. The poor are perceived to be objectified by the programme and used for purposes they are unaware of. For some upper-class audiences, going to the media is an act of desperation that is judged on notions of respectability particularly, shame and honour.

NATASHA: Oh my God. I hate this. They're just so ready to say how miserable they are. So cheap! Have some dignity!

HAROLD: Poor people have no shame, do they?

GLENN: I hate it when other people play up their poverty. It's gross in *Wowowee*. It's gross in real life. Even I was a victim of this injustice. You know, I was supposed to graduate valedictorian in [the university]. But then they selected this guy whose dad died as an [overseas worker] [...] During his [graduation] speech, I told myself, at least I wasn't picked because I wasn't smart enough, it's because I wasn't poor enough.

(Group interview – private university)

Whilst upper-class respondents understand them to be victims, this understanding generates varied feelings. Some express discourses of blame-filled compassion, such as when some hope that contestants win big prizes because of the shame they are subjected to by the programme ('For the added shame that they have to go through, I really hope they win one million pesos!'). Some refer back to the *Wowowee* stampede and demand 'true justice' by having the show's real profits shared with the victims' families. This blame-filled compassion is unique in that it blames not the perpetrators of suffering, as in the descriptions of Höijer (2004) and Boltanski (1999), but the perpetrators of its publicity and televised resolution: the media themselves.

For lower-class respondents, in contrast, *Wowowee*'s participants are understood very differently. They see them as actually possessing a degree of agency. As we have seen, the contestants they watch on the show are people that they pay close attention to and engage with. They become invested in their lives and retell contestants' stories in relation to their own. Contestants' interviews, for them, reveal contestants' moral values and personal worth, which they in turn judge as deserving or not of symbolic recognition and material reward.

But perhaps more significantly, agency is perceived in the very act of going to the media and making a 'media pilgrimage' (Couldry 2003) to *Wowowee*. Such an act is understood as a response to their current situation and as a real chance at making a difference in their lives.

JONATHAN: So you're okay with the format of having the people narrate their stories and then, a minute later, have them dance or even made fun of by Willie?

RED: Oh yes. It's a win-win situation. For the media they get on the good side of the audience. Good ratings. And for the people, it's their big chance! See – you just cry a little and you get some cash. You play some more, then you might even win 100,000 pesos. It's better odds than buying a lottery ticket.

JONATHAN: What about you? Have you thought of joining? You know, like before, when your grandmother was sick? You didn't think of lining up?

RED: Maybe. In fact, my cousin has been encouraging me to join *The Singing Bee* because I know a lot of songs. Soon, maybe. But we're okay in the meantime.

(Group interview –Mandaluyong)

For lower- and some middle-class audiences, the very act of queuing up a night before the show just to enter the mediated centre is paradoxically an indicator of both agency and victimhood of a person. For one, they believe that to queue up is to actively take a chance. There is agency in that it involves calculation, thinking about the odds of winning in comparison to how much money has been won that week, and not simply a whimsical fatalism that leaves victory all up to God. For lower-class Filipinos, there is indeed some agency in *pakikipagsapalaran* (jumping at the chance), as success is understood not purely driven by divine providence but also by human choice. Making an effort to leave one's house is a form of resistance to the physical immobility forced upon them by poverty; 'being at the right place at the right time' in the mediated centre, therefore, is not simply about fatalism, chance, or 'being there [...] doing nothing' (De Jesus 2011), but entails physical and emotional labour. In their consumption of media, audiences have seen the material rewards that symbolic recognition confers. Appearing in the news, being interviewed in a charity appeal and participating in *Wowowee* are all equal opportunities for fame and fortune available in the mediated centre.

To queue or cry or sing or dance or flirt, I learned, are merely perceived to be instrumental for higher rewards. Rather than be consumed by the shame of sharing intimate stories of suffering in public or the laziness of staying at home instead of doing something about one's condition, to play the game and to demand pity from wealthy others are seen as actions that deliver real material consequences, especially when conducted in the mediated centre long known to be hospitable to them. In a way, this is reminiscent of Das' (1995) work reviewed in chapter one on performances of mourning in India being given a 'private vocabulary' by victims of the Partition Riot as a way to demand attention from those in power.

Red, a thoughtful computer repairman, captures it best with his use of the phrase 'win-win situation' in the excerpt above. Lower-class audiences see other poor people in a more or less reciprocal relationship with the media. In my interviews, I have in fact heard the word *gamitan* quite often. *Gamitan* literally means 'making use of each other', and to me it encapsulates how *Wowowee*'s devotees see themselves in relation to the show.

In the space of *Wowowee*, we then find a contemporary and mediated expression of poor Filipinos' strategies of gaining power, demanding recognition and *disponir* (borrowing money) (Cannell 1999). Contestants evoke pity and material reward by managing their appearance and conveying their moral worth through their stories of hardship. Just as Kenney and Clairmont (2009) describe victimhood as being used as a 'sword', I argue that its fans see *Wowowee* as a space of strategic suffering, whereby poor people's assertions of victimhood *and* agency are granted recognition and reward in the mediated centre.

Inspired by Spivak's (1988) term 'strategic essentialism', I use strategic suffering here to describe the creative manner by which people narrate their hardships in order to gain pity, which is then alchemically translated to televised publicity, cash prizes and audience compassion. Strategic suffering involves a mode of self-presentation that goes against typical strategies of connoting high status. Rather it entails a performance that capitalizes on others' sentiments and indignation toward suffering. Strategic suffering could be seen in people's choices of wearing particular types of clothing, ways of storytelling (including the strategic expression of reluctance) and playing up to the affinities of the host and audience.

Strategic suffering can also broadly apply to general acts of borrowing money and evoking pity in everyday life. But when situated in the mediated centre, strategic suffering additionally requires literacy about specific generic codes and conventions – an understanding of the rules of the game. In the context of a confessional interview, it involves deployment of generic and media literacy by knowing when to speak and when to hold back, when to use sorrow and when to use sentiment. It requires the management of one's emotions and skillful narration of one's life story that is consistent with *Wowowee*'s own moral codes and criteria in judging authenticity and deservedness – criteria, which over the years, have become established, even restrictive in favouring only particular claims to suffering.

Fundamentally, strategic suffering is underpinned by a particular understanding of a person's relationship with the media: the media are seen here not as distant or exploitative but in fact intimately dependent on the *masa* that they claim to serve. Lower-class audiences know that the drama and entertainment of suffering are valuable commodities to media institutions, so the ability to manage one's pitifulness whereby one might be seen as deserving (of compassion, recognition and redistribution) is recognized as a worthwhile skill.

Once again, middle-class respondents have much more plural interpretations about the agency of sufferers in the media. What seems to be common is their seeking of resourcefulness and hard work among *Wowowee* contestants. Whilst they evaluate the programme as possibly encouraging false hopes of personal transformation and false values of dependence on rich patrons, they are also sympathetic to certain contestants who exhibit agency not only in the act of queuing for *Wowowee*, but also in their personal stories of coping with their conditions of poverty.

Play and Pity: Ecological Ethics of *Wowowee*?

This chapter provided an empirical account of people's responses to suffering in the context of a popular and controversial entertainment programme. By examining the moment of reception, we saw how different discourses of compassion are expressed, how these discourses were related to people's class and cultural contexts and how these discourses were informed by moral judgments of sufferers and the media that play host to sufferers.

Also crucial here were culturally specific understandings of agency that take into account varying conceptions of what true suffering means and how it appears, and the moral qualities that sufferers must possess to deserve compassion, recognition and redistribution over others. We found that the agency of sufferers was not solely a property of their textual representation, but is also a subjective judgment of audiences.

This mapping out of audience ethics (see Table 2) in the reception of a TV programme once again signals the social – and ethical – problem in how upper-class Filipinos perceive less privileged others in society. They judge the poor as a homogeneous mass with low moral value in their thoughtless participation in the shameful public space of *Wowowee*. Such judgments lack acknowledgment not only of poor people's individual personhood and their own skills and literacies to read and follow media codes and conventions; they also lack acknowledgment of structural conditions of inequality that push them to perform suffering in excess on television. Elite assumptions that the poor queue up in the mediated centre to simply 'be there' and 'do nothing' (De Jesus 2011) or willingly shame themselves for money are proven to be unfounded in this study, as queuing in the mediated centre and narrativizing suffering in public constitute real and difficult forms of physical *and* emotional labour for the poor. Queuing entails calculated risk and can even be an act of desperation when other avenues for help have betrayed them. Narrativizing suffering on television is ultimately a sacrifice of their personhood, as they subject their life choices and actual physical bodies for the scrutiny and judgment of potential patrons and judgmental audiences. Typically, upper-class statements of disgust about *Wowowee* and its audience are informed by stereotypes of the poor as well as a lack of media/generic literacy of accepted codes of conduct in this space. Upper-class audiences' lay media moralities, the term I use to refer to

Table 2: Mapping Moral Responses to *Wowowee*

Audiences	Consumption of *Wowowee*	Discourse of Compassion	Judgment on Sufferer	Judgment on Media-Sufferer Relationship	Judgment on Genre-Sufferer Relationship	Criteria of Deservedness
Upper class	'Switching off'	* Usually none * Some blame-filled compassion (blame directed not to perpetrators of suffering but to media that publicize suffering)	Victim (homogeneously poor; unwittingly exploited by media)	Highly asymmetrical * exploitation * mystification * shaming	Play and pity Contradictory	Respectability * vs shame of exposure * vs shame of being exploited by others
Middle class	Occasional viewing (some older people were critical of the show's charitable practices)	* Some tender-hearted compassion (directed to individual sufferers) * Some blame-filled compassion (directed to media that perpetuates false hopes and values of dependency)	Variable (agency possessed by deserving, hardworking few)	Asymmetrical * media perpetuating wrong values of fatalism and dependency to the poor	Variable * acknowledges need to reward and recognize resourceful and hardworking sufferers	Resourcefulness and hard work * vs overreliance on rich patrons * vs overreliance on games of luck * vs overreliance on media and other institutions

Lower class	Affective consumption (some men were critical of contestants' overt displays of emotion)	Tender-hearted compassion (directed to individual sufferers)	Agent (agency evidenced by mere act of queuing for *Wowowee* and taking a chance/pakikipagsa-palaran)	Minimally asymmetrical or reciprocal * win-win * dependency * last resort at times of need * Willie as rich uncle	Play and pity Compatible	1. Truth * material indicators of poverty * subjective indicators of authenticity 2. Fairness * respect for elders, fulfilling obligations to family, resourcefulness * commensurability of prize to recipient

audiences' judgments of good and bad media conduct, were revealed in this chapter to be informed more by speculations about media content and media audiences than by sustained attention or understanding.

In upper-class respondents' everyday decisions to 'switch off' from public displays of suffering on *Wowowee*, they 'fly over' (Tadiar 2004) and resist the obligation to recognize the poor and engage with the realities of their life conditions. Such perceptions and practices of the upper-class towards their less privileged others signal a relationship that lacks proper distance, and is 'too far' (Silverstone 2007, 48) – both from their representations on television and their actual appearance in everyday life. In other words, the upper-class evaluation of complete victimhood in *Wowowee* ultimately reveals more about a general disengagement towards suffering others rather than an empathetic understanding of them and their media. The question to be asked here then is: How valid are cultural elites' criticisms of *Wowowee* and its audiences when such media criticism is characterized by knee-jerk disgust, limited knowledge of local media conventions, and general disinterest with the cultural practices of the *masa*? I return to this crucial question in the final chapter.

As much as lay moralities have been the focus of this chapter, the ecological ethics of *Wowowee* must now be addressed. On the one hand, we have indeed found through this case study that it is difficult to provide a definite resolution to the long-running media ethics quandary of whether to portray sufferers as humane or empowered (Tester 2001) or 'at [their] worst' (Cohen 2001, 183; Orgad 2008, 21), given that audiences' discourses of compassion are informed by different and competing evaluations of agency. Further, we have seen that a particular textual moment of suffering, regardless of how it represents the agency of a sufferer, may not productively evoke any discourse of compassion anyway if an audience member had already switched off from the media to begin with.

On the other hand, I argue however that it is possible to provide criticism of *Wowowee* by focusing on the kind of agency that *Wowowee* values and legitimizes as well as by identifying the social consequences that the show has in wider public life.

Undoubtedly, the bottom-up perspective of this study attests to how *Wowowee* functions as a legitimate space of recognition and reward for people historically and actively refused these. *Wowowee* fills the gap left by legal or public institutions in which claims to injury and victimhood can be made. Parallels can be drawn with how strategic suffering is performed in a similar way that Indian women found alternative places to mourn when denied legal recognition by the government (Das 1995). *Wowowee*'s 'unveiling' of the other (Silverstone 2007) and willingness to 'give voice' to the voiceless (Cottle 2006) are practices that are experienced by my low-income audiences in stark relief to their everyday feelings of being excluded, ignored and de-valued in public life.

However, giving voice and dispensing help aside, the kind of agency that *Wowowee* encourages can be criticized. *Wowowee*, as a programme that follows conventions of reality television and talk show, is surely subject to existing criticism of these genres, which according to their critics operate under the broader ethical regime of neoliberalism (Couldry 2010; Illouz 2003; Orgad 2009; Wood and Skeggs 2009). Such a regime promotes 'governmentality, competition and elimination' (Orgad 2009, 156) and universalizes values of individualism and self-responsibility by forcing the working class to transform themselves (Wood and Skeggs 2009, 187). Similarly, *Wowowee* imposes challenges of self-transformation on its poor contestants through clichéd encouragements to 'work hard' and 'continue being good persons' with no acknowledgment of the economic, political or cultural forces in society that keep them poor.

However, unlike in the West, where occupying a position of low value is blamed solely on the individuals and their lack of capacity to transform themselves (Skeggs 2004), here poverty is also acknowledged as a fact of life and is partly produced by divine/supernatural forces (*'Hay buhay'*/'Such is life', as Willie was quoted in one of the excerpts above) – reflective of Catholic beliefs shared by many in the country. *Wowowee*'s moral discourses of suffering then provide coping mechanisms for the poor in its underlying message that there are limits to their personal agency in a world also influenced by divine logic. But as David (2001[1976], 42) says, the language of coping mechanisms provides little invitation to seek root causes of social problems. For a programme that claims that it exists to help the poor, perhaps we must demand that *Wowowee* make better use of its platform than to rehearse routine pronouncements of fatalism that reinforce rather than challenge the status quo.

We must also criticize *Wowowee* for legitimizing only certain kinds of suffering over others. Rather than seeing *Wowowee* as a democratic space where minority voices are heard, *Wowowee*'s conventions only authorize highly specific claims for recognition and reward, whilst actively de-valuing others. *Wowowee* prefers individual rather than collective testimonies of suffering, emotional/pity-based claims rather than rational/justice-based claims of suffering and disclosure rather than silence in bearing witness to one's suffering. Whilst its legitimization of individual, pity-based and full-disclosure performances of suffering may be effective in gaining the affection of its poor audiences, it does little to inform them about equally legitimate forms of suffering and claim-making. Authorizing alternative claims to suffering can educate them about possible claims that they can also make on the government, legal institutions, unions and employers about injustices that they experience either individually or collectively. From my interviews, I have indeed found that, for example, lower-class respondents have limited knowledge of how to access certain services from government institutions or seek out reparations or solutions to job-related grievances.

In addition, democratizing the *Wowowee* space to accept alternative (rational/ justice-based and low-disclosure) claims to suffering may widen its appeal to upper- and middle-class audiences, who may have the political and economic resources to address particular claims to victimhood. In *Wowowee*'s over- representation (Wood and Skeggs 2009, 177) of only certain kinds of suffering and certain kinds of people, this mediated space ultimately contributes to and amplifies class conflict and misunderstanding in a society historically marred by various economic, political and cultural divides. The challenge perhaps for television networks today is to find ways in which the genres of reality and talk show can represent various kinds of sufferers that may invite discourses of compassion across social classes. To paraphrase Orgad (2009, 156), a democratic and moral mediated space should not allow only the 'fittest' and most televisually entertaining claims to suffering to survive.

Finally, by addressing claims of suffering of particular individuals whom the host and the network judge as deserving, *Wowowee* perpetuates personalistic patron-client ties, whereby poor people continue to depend on rich patrons for protection and salvation. This arrangement ultimately benefits *Wowowee* and its television network ABS-CBN. Just as power typically works in systems of gift exchange (Mauss 1966), the patron ABS-CBN accrues power not only by the amount of resources that it acquires but also by the amount of money that it gives to less fortunate others. Although *Wowowee*/ABS-CBN's actual income comes from its corporate sponsors, its practice of extending its reach by fashioning itself as an economic mediator in this context allows it to convert capital from corporate sponsorship to compassionate charity for 'deserving' individuals. These transactions and conversions of capital from one form to another are processes that are conveniently obscured to those outside the mediated centre. And so, in the end, through economic mediation and symbolic substitution, the television network stands in for the sponsors, donors and public actors, who may have been instrumental for charity to have been enacted, and strategically accrues for itself a trusting and loyal viewing audience.

Whilst *Wowowee* may evoke different discourses of compassion from audiences based on varying interpretations of the agency, this mediated space is still morally culpable for the processes by which it interacts with the sufferers it plays host to. On this level of analysis, I argue that the quantity of the reward and recognition it grants to the 'deserving' are ultimately incommensurate with the quality by which people's beliefs, actions and moral worth come to be undemocratically represented and regulated within this mediated moral space.

Chapter 5

NEWS:
RECOGNIZING CALLS TO ACTION

People who are remote, socially or geographically, can afford to be
unhelpful. Their reputation in the eyes of those from whom they are distant
is of little to no importance to them.

Benedict Kerkvliet, *Everyday Politics in the Philippines*

Drawing primarily from focus group and life story interviews with audiences
and supplemented by expert interviews with media industry representatives,
this chapter describes the contours of audiences' engagement with the
news in relation to media ethics debates. News audiences, particularly their
capacities to act either as engaged publics or disinterested bystanders, have
been the traditional subjects of audience studies on compassion fatigue
(Höijer 2004; Kinnick et al. 1996; Kyriakidou 2005, 2008), whilst news texts
have been the focal point of arguments about the best and most ethical ways
of representing suffering. Whilst the previous chapter presented audience
responses to suffering in the genre of entertainment television and challenged
some concepts and assumptions made in the news-centric media ethics
literature, this chapter presents material in direct dialogue with existing work
in the field.

This chapter finds that the news is a complex genre that offers highly
varied representations of suffering. Unlike in *Wowowee*, where programme
conventions recognize and reward highly specific claims to victimhood in
participants' 'strategic suffering', the news offers *diverse* forms of suffering:
nearby and distant, natural disaster and man-made, individual and collective,
pity-based claims and justice-based claims. And again, unlike in *Wowowee*,
which only involved poor people visiting the media centre, the news also (and
primarily) involves 'reverse pilgrimages' (Couldry 2003, 93), where media
people visit ordinary people in their everyday life contexts. By having plural
representations of suffering and multidirectional media pilgrimages, the news

provokes diverse lay moralities of suffering and lay media moralities of good and bad media conduct in relation to the 'Poverty of Television', which in the Philippine context involves media intervening in social affairs in the context of a weak state.

At the same time, this chapter uncovers important continuities in audience engagement across entertainment and news genres. Classed consumption patterns of 'switching off' and affective consumption are likewise observed here. Similarly, moralities of authenticity and deservedness are employed in evaluating televised suffering. I find that, in the context of news, these moralities have consequences that affect audiences' decisions to act on particular moments of suffering. Unlike in entertainment, talking about suffering in the news involves audiences *working through* questions of action, as was evident in the context of a research interview. Unlike in entertainment, my news interviews contained qualities of the 'moral dilemma interview' (see chapter one), whereby moral reflections on and justifications of action (and inaction) are articulated. Indeed, the news genre enjoys greater trust from audiences and wider acceptance of its role in addressing important public issues (Livingstone 2005). Because of Philippine news conventions that embed charity appeals within the newscast, audiences could not but provide direct acknowledgement, or outright rejection, of invitations to help suffering others during interviews.

This chapter begins with a brief review of existing scholarship on the news genre. The second section maps out dominant beliefs and practices in the Philippine news industry. The third section takes us through an ethnographic exploration of audiences' patterns of news consumption and suggests how these patterns are indicative of people's understandings of suffering and their obligations toward sufferers in the news. Then I discuss a case study of audience reception of distant suffering, where we find more similar than different responses across different groups of audiences. The final section reflects on several cases of audiences' charitable practices that were triggered partly by the news and partly by the audiences' social contexts.

A key finding of this chapter is that news and entertainment are similar in that they both 'overrepresent' (Wood and Skeggs 2009, 177) local sufferers and therefore offer moral claims for audiences to express discourses of compassion. However, a crucial difference between the two genres is that television news provides moral claims of more *diverse qualities* than the entertainment genre: representations of suffering with varying visual codes and emotional intensities that audiences can variably engage with. In this way, the news occasionally enables audience actions such as charity and volunteerism, which exceed modest requirements of 'effective speech' (Boltanski 1999, 18–19).

News and Normativities

Whilst there are similar concerns about narrative ethics and representations of the other in fictional media genres in the humanities (Attridge 2004; Gibson 1999; Onega 2010; Pandya 2011), the media ethics literature has paid considerable attention to the factual media genre of television news. News consumption is considered a mediated context that offers possibilities of audience ethical responses, such as speaking out about the suffering of others (Boltanski 1999), challenging stereotypical representations of the other (Silverstone 2002, 762) and donating to humanitarian organizations (Chouliaraki 2006; Tester 2001). The media practice of tuning in rather than 'switching off' from the news is widely regarded as a democratic (Jensen 1995; Putnam 2000) and moral (Cohen 2001; Seu 2003) duty across the disciplines of political science, sociology and media studies. Media ethnographies have also uncovered that audiences themselves express positive value judgments about news consumption, associating values of maturity, self-efficacy, social belonging and citizenship with this practice (Buckingham 2000; Gillespie 1995; Madianou 2005b).

The less normative strand of Silverstone's research contributes a different perspective to this discussion by stressing more modest and instrumental functions of news in the conduct of everyday life (Silverstone 1994, 2005). Developing Giddens' (1991, 38) notion of 'ontological security', Silverstone (2005, 196) argues that television news can foster feelings of constancy and confidence in the stability of the world through its 'paradoxical role in the massaging and managing of collective anxiety'. Silverstone highlights how the time scheduling of news to coincide with domestic patterns (e.g., breakfast news, lunchtime news, etc.) and news anchor practices of 'mutual smiles and silent chat following a "human interest story"', among other genre conventions, provide a sense of reliability and control amidst stories of seemingly uncontrollable catastrophe (Silverstone 1994, 16–17). In the context of televised suffering under investigation here, I intend to probe how news consumption may be a contested site in which invitations to engage with public issues contend with more personal and instrumental uses of the medium. In addition, the idea of ontological security relates with arguments about Filipino coping mechanisms, as we reviewed in chapter one. Bankoff's (2003) 'culture of disaster' thesis suggests that Filipinos respond to suffering not with shock and panic but with adaptive fatalism and anxiety management. Seeing how disasters are domesticated through news consumption and everyday talk may lead us to see any cultural specificities (and commonalities) in people's responses to suffering.

We can also tease out from Madianou's (2005b) research that scholars' normative judgments about news audiences may not capture the diverse experiences of audiences in their everyday lives. For instance, her ethnography discovers that habits of 'switching off' from the news were actually active responses by ethnic minorities to their systemic misrepresentation and exclusion from Greek public life (67–72). This finding suggests that it is important to carefully contextualize audiences' decisions to 'switch off' before pronouncing normative judgment of right and wrong about this practice. From previous chapters, we have seen how upper-class respondents switched off from local media, partly from media conventions of 'overrepresenting' (Wood and Skeggs 2009, 177) the poor and partly from everyday habits of 'flying over' (Tadiar 2004) the realities of suffering around them. This chapter intends to probe whether these reasons similarly apply to the news.

Finally, following insights from ecological ethics approaches, Couldry's (2003) notion of 'media pilgrimages' and Madianou's (2005b) discussion of people's 'direct experiences with the media' are useful to review in the context of news. Just as we saw in the previous chapter, audiences' discourses of compassion toward sufferers in the media are also shaped by their evaluation of media pilgrimages. Since news affords diverse and multidirectional media pilgrimages and reverse pilgrimages, it is fruitful to explore then the conditions of media pilgrimages that prompted audiences to not only utter discourses of compassion but also donate or volunteer in aid of nearby – and perhaps even distant – suffering others.

Sleuths and Saviors

ABS-CBN's and GMA's primetime newscasts are among the top ten highest-rated programmes on Philippine television (McCann Erickson 2009). ABS-CBN's *TV Patrol World* (Mondays–Fridays, half past six to eight o'clock in the evening; Saturdays, half past six to seven o'clock in the evening; 1987–present) and GMA's *24 Oras* (24 hours) (Mondays–Fridays, half past six to eight o'clock in the evening; Saturdays, quarter past seven to half past eight in the evening; Sundays six o'clock to quarter past six in the evening; 2004–present) consistently rank alongside the most popular soap operas and factual entertainment shows in total day ratings, each typically garnering a range of 26 to 35 per cent of total viewer households.

Apart from these, the two networks produce early-morning news segments, hourly news bulletins and full-length late-night newscasts. ABS-CBN has a 24-hour English-language cable news channel called ANC and is among the highest rated cable channels (McCann Erickson 2009). More recently in 2011,

GMA launched its own news channel – the Tagalog-language GMA News TV – on free television.

The news and public affairs departments of ABS-CBN and GMA contribute to the networks' image of trustworthiness and credibility to the Filipino public, garnering for both companies prestigious awards such as 'Asia's Most Respected Company' (Rodrigo 2006, 389) and 'Reader's Digest Gold Award for Most Trusted Brands' (GMA News Online 2010). The most cited evidence of viewers' trust toward network journalists and personalities is perhaps the great number of them elected to public office (Coronel 1999; David 2001c[2000]). News anchors have reached positions of senator and vice president. Public affairs programmes helped catapult their hosts to positions such as Mayor of Manila (e.g., Fred Lim, a former policeman). Today's politicians have since joined television as programme hosts to capitalize on the status of television as 'kingmaker' – a term accurately, if jealously, coined by print journalists (Coronel 1999).

This high visibility of and trust toward journalists and news personalities are in part due to the storied legacy of journalism in the country, as mentioned in chapter two. Members of the Filipino intelligentsia who published newspapers and pamphlets to spread their intimations of revolution against the Spanish Empire in the 1890s were later declared national heroes. The profession of journalism is intimately linked to ideas of nationalism and patriotism in the Philippine context, suggested too in Anderson's *Imagined Communities* (1983, 26–9).

Previous studies on Philippine journalism have also focused on its consistent spirit of anti-authoritarianism and resistance during both colonial rule (Maslog 1990) and Ferdinand Marcos' martial law regime (1972–86) (Del Mundo 1993; Pineda-Ofreneo 1986). The post-martial law 1987 Constitution, used to this day, serves as witness to the trauma of the period, establishing multiple parameters for checks and balances intended to prevent any future curtailment of journalistic freedom by totalitarian rule (Smith 2000). After martial law, television was 'awash with public affairs talk shows' since the 'people were hungry for information after 14 years of censorship' (Stuart Santiago 2011). But in the 1990s, business concerns began to weigh on network owners' minds. ABS-CBN made an earlier decision to be a 'mass-oriented station' and enjoyed early dominance of the television landscape over its rival, GMA (Romualdez 1999, 55). The shift of local newscasts from English to Tagalog most explicitly signaled this transition to mass-oriented programming. Whilst David (2001c [2000], 147) thinks that this shift was 'marvelous' in making political affairs accessible to the wider public, he also laments why this coincided with 'a distinctly showbiz vocabulary': 'it's almost as if Tagalog is unable to shake off its close affinity with movie world gossip and scandal'.

Today, there is much criticism of television news and their 'tabloidization' (Rimban 1999) and 'dumbing down' (Contreras 2011; Stuart Santiago 2011). Even journalists that I interviewed for this study expressed ambivalence about the pressure to 'appease the *masa*' and produce programmes with 'always the *masa*' as the intended audience. GMA Public Affairs Programme Manager Angeli Atienza shared with me that their team's shorthand to decide on programme topics and story angles is expressed in the question '*Mataas o masa?*' (High/highbrow or mass?). *Mataas* here can refer to topics, such as international politics, stock market and policy debates. Low-emotion and facts-based ways of reporting are also labeled as *mataas*; Atienza shares the assumption that *mataas* issues always 'need a face', or personalization, to ground the story in the everyday lives of their target audience.

From my interviews with journalists, 'the *masa*' is also cited as a frequent reason for the local and parochial quality of news programmes. For instance, world news is covered in a two-minute segment called 'World Patrol' in the curiously titled *TV Patrol World*. Here natural disasters, political conflicts, celebrity news and human-interest stories across continents are breathlessly reviewed altogether. Certain international news items are sometimes granted wider coverage when these items might impact Filipinos in the country or overseas.[1] For example, the June 2009 Air France plane crash in Brazil garnered headline status in television and print with a specific concern for the one Filipino among the 200 passengers. The general impression of news producers is that world news is not relevant for their *masa* audience, as their primary concern remains providing for their families.

From my interviews with journalists, I nevertheless identified a shared belief that the best journalists are those who balance *mataas* with *masa*. Producing well-produced and populist news and public affairs programming is interpreted as in fact the best and most pragmatic form of public service, since the *masa* are understood to be the sector most in need of education, entertainment and charity. In other words, the tension between serving the market and doing public good is not always conceived as contradictory in the Philippine context. As former GMA executive producer and section editor of newspaper *Manila Bulletin* Melo Esguerra told me,

> Really, the duty of the Filipino journalist is to serve the *masa*. Because the *masa* has really nowhere else to turn to. They do not have the government to turn to because the government has no money for them. In fact [journalists] are expected to take on the role of savior [...] We have no bureaucracy. Our help is instant.

Esguerra here highlights another peculiar feature of Philippine news: the formalized synergy between news and public affairs departments and the charity organizations of the networks. As mentioned in previous chapters, ABS-CBN and GMA Network both have charity organizations that work alongside journalists and news producers. Aside from natural disaster assistance, other 'reverse pilgrimages' involve provisions of free medical and dental check-ups, free circumcision operations, free legal consultations, free health insurance inquiries and free wheelchairs during visits to slums and rural communities. Typically these activities coincide with anniversary episodes of public affairs programmes, where hosts claim that they intend to 'thank' and 'give back' to their loyal viewers. Sometimes, they are part of campaign-style road shows to strategically promote the network in cities significant for ratings tallies and corporate sponsor interests. Such events can generate attendance of as many as 60,000 people (Martinez-Belen 2009). Television network research suggests that public service events foster viewer loyalty, thus preventing channel-switching (GMA Research 2004).

People visiting the media centre meanwhile access similar provisions of legal advice, medical assistance and referrals to NGOs and other government offices – albeit at a smaller scale than the special reverse pilgrimages. In certain provinces in close ratings competition, networks run microfinance programmes that provide special moneylending services to poor people. In Iloilo and Bacolod, debtors can borrow up to 5,000 pesos (70 pounds) on the condition that the debtor's television is consistently tuned in (to ABS-CBN). From my interviews with network executives, I learned that a roving motorcycled monitor regularly checks up on debtors and demands immediate repayment from those caught watching the rival network.

Within the evening primetime newscasts, there are also special one-to two-minute segments that feature charity appeals and network charity activities. In GMA for example, *24 Oras* news anchor Mel Tiangco also acts as Chief Operating Officer of their charity, the Kapuso Foundation. Within *24 Oras*, a segment called 'Kapusong Totoo' (Truly one of heart) airs charity appeals. The charity appeal employs documentary-style generic techniques where the mise-en-scène is the everyday, usually domestic, context of a poor person struggling with a grave illness. There is a 'curiosity of the flesh' in the camerawork as close-ups focus on the subject's frail figure, her/his passivity laid bare for viewers to witness. The appeal includes a doctor's diagnosis of the patient's illness and explains the cost and urgency of the operation. Typically, the mother of the ill patient makes a direct appeal to viewers to help her ill child, testifying to her inability to provide for her child's medical operation. Then the segment cuts to the studio, where Tiangco makes a live appeal to individuals or corporations to donate to the cause. The Kapuso Foundation telephone number is flashed, and Tiangco reminds viewers that they take calls from donors as well as other sufferers in the audience requesting for similar kind of assistance. From my interviews with Kapuso Foundation personnel, they say that the poor person on television is typically selected from people who had visited their offices to make direct appeals for help (for a longer discussion, see Ong 2015a).

Philippine news media then do not commit social denial about poverty, as critics have observed about media in India (Mankekar 1999; Sainath 2009), as mentioned in the introductory chapter. Whilst media's provision of social services to citizens is also observed in countries with weak state structures for welfare such as Mexico (Garcia Canclini 2001; Pertierra 2010), peculiar to Philippine media is the institutionalization of the news/charity link. But whilst this practice is coated in the discourse of public service by the media themselves, the business motives behind such practices are evident to critics.

Abuses of the high level of trust and power enjoyed by news media have also been documented. For instance, Hofileña's (2004) *News for Sale* served as an exposé of journalists accepting bribes from politicians to act as covert public relations mouthpieces. In August 2010, there was also extensive public criticism of ABS-CBN's and GMA's news coverage of a hostage crisis in Manila that ended up killing eight Hong Kong nationals. The live coverage was said to have contributed to the failure of the rescue mission, as the hostage-taker tracked the movements of the police via the live news broadcast on the television inside the bus where he held the hostages. Journalism colleges issued a public statement decrying how the live coverage ruined the 'element of surprise' for the police to capture the hostage-taker and was ultimately motivated by networks' 'competition for higher ratings' and 'exclusive' reports (Tolentino 2010). In my own work, I have also documented how journalists deflected all criticism of their coverage by collectively directing critique at the government (Ong 2010).

In the next section, we explore audiences' patterns of tuning in and 'switching off' from the news. Drawing from life story interviews and participant observation, I highlight the different uses of news in people's everyday lives and suggest what these might mean for their interpretations of suffering and obligations towards suffering others.

Regimes of (In)Attention: Security and Anxiety in News Consumption
Lower-class audiences

It was a humid Monday evening in early June 2009 when I visited my informant Amparo at her home in Park 7. It was my first time visiting her house, as our previous chats were held in the more easily accessible basketball court, which functions as a makeshift town plaza where neighbours meet and gossip. I was a few minutes late for our scheduled *TV Patrol* viewing; I had lost a half hour going back and forth narrow, serpentine alleys looking for Malunggay Street, stopping every so often for directions. When I arrived, Amparo, a lively, confident woman approaching her forties, was quick to say, 'Oh! Come in, come in! You might miss the latest!' As Amparo helped me up the wooden ladder to her home, I had to blink my eyes a few times to help adjust from the dim outdoors to what was unmistakably the bright white glow from the television in the centre of her home. We sat on the floor and started watching the news.

That evening in June, there was a news report about the first week of classes and conflicts between government and local state schools. The story discussed how elementary school principals were rejecting pupils because their classrooms were already packed with ninety students each. Interviews with officials were interspersed with images of small classrooms, children elbow-to-elbow sharing textbooks and tearful mothers making appeals for help. A few seconds of the report were devoted to a five-year-old boy standing and sobbing in a corner; the journalist explained that the boy was frustrated that he had nowhere to sit in the overcrowded classroom. Footage of the teacher determinedly lecturing inside a cramped, chaotic room closed the story.

After this, Amparo turned to me, her eyes blinking, as if holding back tears, 'I just feel sad for that boy. I just remembered my childhood [...] I remember how I told myself that I would not allow that to happen to my daughter.' As other news clips played, Amparo continued talking about this report and her personal disappointment with not making it past fifth grade. I tried to console her, 'It's okay, it's okay.' She said, 'Yes. I know it will be okay. In fact, when I watch the news, I gain inner strength (*lumalakas ang loob*). I will not let that happen to my family, I say. And I remember I am so blessed by God that our family is complete, that we have food to eat.' She cited the previous week's news about a capsized passenger boat where several children had perished. When she watched that, and other stories of that kind, she said that she realized how 'lucky' she actually was, how their family was 'still blessed by God'.

Amparo's melodramatic approach to her life is in a way mirrored by her affective consumption of the news. Just as we saw viewers of *Wowowee* gain a sense of hope and security by comparing their personal situations with the more pitiful plights of others, Amparo and other lower-class respondents draw resources for coping from the news. They reflect on how the conditions of people featured in individual news clips might compare to their own. As mentioned earlier, local conventions of 'personalizing' public issues through the perspective of the poor encourage viewer identification. Local news conventions of portraying 'private' scenes of mourning (i.e., especially of family members of the victims covered by news stories) and depicting graphic images of accidents and disaster, whilst heavily criticized by cultural elites such as Carlos Celdran (in Santiago 2009), provoke emotional responses of compassion and indignation from lower-class audiences.

Just as lower-class respondents engaged with contestants in *Wowowee* and found resonances with sufferers' life stories, the news likewise becomes a similar occasion for *damay* (sympathy/mourning), whereby feelings of sorrow, (self-)pity, blessedness and enduring hope are expressed. News is retold not only as discrete events or isolated stories, but also in referential and personal ways, where individual victims of tragedy stand in for many other poor people like themselves. Though previous studies have also uncovered how populist news helps people 'cope with their lives', here it is less about the news being an 'endless source of laughs' (Bird 1992, 204–5) than news engendering among viewers thoughtful reflections and comparisons between grave and graver realities of suffering.

From my lower-class interviews, I found little evidence of textbook 'compassion fatigue', if we are to understand it as avoidance of televised suffering or desensitization to repeated images of suffering (Cohen 2001, 190–1). Whilst there is indeed repetition – what with the natural disasters and deaths depicted on a daily basis – this repetition is experienced by the lower class as a *continuation* of a larger story of suffering that unites them with an imagined moral community with other poor sufferers. Whilst there is a sense that the 'story is the same' (that life is still hard) and 'only faces change' (that new characters are substituted for timeless roles of villains or victims), their story of 'patient endurance' (*pagdaos; pagtitiis*) in the face of suffering continues. Contrary to studies of compassion fatigue that argue that the news fosters feelings of helplessness among audiences (Höijer 2004, 523), I observe that viewers can actually derive a sense of personal agency for themselves from being made aware of *others'* experiences of suffering.

Additionally, lower-class audiences provide the occasional judgment that sufferers in the news are *undeserving* of either news visibility or significant attention from audiences. This judgment is given when they assess that the news subject's condition is less grave than their own, more 'authentic' condition of suffering. An important idea here is that, similar to their expectations of *Wowowee* producers, they expect news editors and journalists to prioritize 'authentic' sufferers who are truly deserving of news visibility, particularly the suffering Filipino poor. I remember here the forty-two year-old laundrywoman Elsa, who criticized a news report about a fire that ravaged a middle-class subdivision; 'surely,' she said, 'there could have been graver events' covered by the news.

Upper-class audiences

As we saw in previous chapters, upper-class audiences tend to display avoidance strategies of local media in the assumption that these purely cater to *masa* interests. However, unlike with entertainment, completely 'switching off' from the news can never be fully justified, even among exclusively upper-class circles. The linkage between news and ideals of civic duty – observed too in other countries (Jensen 1995; Livingstone 2005; Putnam 2000) –hinders complete withdrawal due to threat of moral sanction. Filipino elites, in particular, have also been observed to claim and maintain their high social status by displaying their knowledge of Philippine public issues as enabled by news consumption (Ong and Cabañes 2011).

The first pattern that I identified is that news consumption is irregular among my upper-class respondents. As discussed in chapter three, they lead mobile lives, and therefore may often miss the primetime evening news; some may still be at work whilst others may be stuck in transit. Some express that they are too tired to stay up and watch the late-night newscasts. They also claim that they lack time to sit through entire newscasts. When my respondents say that they 'follow the news', they actually follow news about discrete events, not complete broadcasts. Following the news often begins with alerts through text messages sent by family and friends, news articles posted on Facebook and recommended news links on their Yahoo! homepage. For example, Steph, a university student, shared with me that her mother is frequently first to alert her 'when news happens', such as terror warnings, bomb scares or kidnappings. These momentous occasions would prompt her to tune in to the news. Because upper-class audiences tune in during exceptional moments, the news clips that they encounter may tend to follow conventions of ecstatic news such as liveness and dramatic language (Chouliaraki 2006, 158–9). So, whilst upper-class audiences recognize that local news is regularly about bad news of disasters and conflict, they still regard the sufferings of others as distinctive events that interrupt daily routines and puncture their psychological zones of safety.

The second pattern that I identified among upper-class audiences is how tuning in to local news is accompanied by experiences of shock and disgust. Compared to lower-class audiences, who in fact seem to gain some reassurance from watching other people's (more) difficult conditions of suffering on television, upper-class audiences tend to register anxiety, even surprise. For instance, when I watched the news with 23-year-old advertising executive

Ronaldo, he reacted differently from Amparo to a similar news clip about overcrowded classrooms.

> RONALDO: Oh wow. I didn't know it was that bad! Jonathan: What was bad?
>
> RONALDO: That they could cram that many students in one classroom. I thought at least that we would have policies for such a thing.
>
> JONATHAN: You've never heard of this before?
>
> RONALDO: Well, kind of. I have an idea, I mean. But then, it's just – wow – shocking, you know? To actually see it.
>
> (Individual interview – upper class)

Like other upper-class respondents, Ronaldo feels saddened and depressed by news items about the experiences of the Filipino poor. Having limited exposure to poverty, upper-class respondents express being 'disturbed' when they have to acknowledge realities that they have otherwise been unable – perhaps even have refused – to see. In this light, they experience less ontological security in news consumption. Though they derive a sense of reassurance from how news confirms that a natural disaster or other bad news does not directly impact their lives, the shocking images in the news provoke worry, doubt and anger. Additionally, unfamiliarity with local news genre conventions of portraying graphic imagery and depicting emotional scenes of mourning along the lines of 'shock effect' representations (Chouliaraki 2010, 111) draws criticism and prompts 'switching off'. Charity appeals that feature children with enlarged heads from hydrocephalus and tearful mothers making direct appeals to the camera are evaluated as 'disturbing'. Upper-class audiences perceive that an excess in both quantity and quality of bad news is represented, and many express a demand for local news media to prioritize positive and uplifting stories – a view also shared by government officials (Bordadora 2011) and 'old rich' cultural critics (Celdran in Santiago 2009), who express disgust with the 'washing of dirty linen in public' (Cordova 2011). This upper-class desire to manage the publicity of conflict and suffering runs contrary to the lower class's agreement with news conventions of prioritizing the poor and graphically depicting their conditions.

As a result, some upper-class respondents either 'switch off' or at least turn to alternative sources of news. Articles in newspapers and news websites are predominantly text-based rather than audio-visual and tend to have less emotional impact. A few consume more international than local news, especially those who have relatives abroad and have plans for travel or migration.

Middle-class audiences

The news consumption of my middle-class respondents finds itself, once again, somewhere between the two poles. I found that my middle-class respondents are less predictable: some watch the news regularly, whilst some tune in only on momentous occasions. Though they are less likely to be shocked by knowledge of issues related to poverty or graphic images of illness or death, they also register great anxiety when watching the news. As computer technician Bumbo told me, 'My heart races when the newscast comes on. I even take a deep breath and try to prepare myself for the bad news that's to come.'

As we have learned from Parreñas (2001) and Pingol (2001), the middle-class position is a precarious position that can be 'lost as a result of external macro- or micro-level events. And so, the news is often switched on with some degree of trepidation that the day's events of disaster might have a direct or trickle-down effect on them. It must be noted that they also experience a sense of ontological security when they realize that they have been spared from disaster. For the most part, though, the middle-class seems to be the most susceptible to dramatic life changes as a result of natural disaster or tragedy. Whereas members of the upper class have social and economic capital to help them maintain or regain comfort and the lower class has few actual resources to muster in moments of tragedy, middle-class people can very swiftly slip from a position of relative ease to a position of great hardship and shame.

This ethnographic section uncovers that news consumption practices are indicative of classed interpretations of suffering. For upper-class respondents, who are shocked and disturbed by discrete news events, suffering is seen as *event*: typhoon, earthquake, educational crisis, terror attack. But for lower-class respondents, who are regular viewers of the news and regard news stories as part of the 'flow' of suffering on television, suffering is seen as story and property of the *poor*. In their eyes, rather than tragedy leading to an experience of suffering, the suffering are in fact *led* to experience tragedy. The Filipino poor, living in zones of danger, are those understood as most vulnerable to the tragic effects of landslides, volcanic eruptions and overloaded transport ferries. The news functions in their lives as a continuing documentation of the everyday struggles of the Filipino poor, rather than individual clips of exceptional tragedy.

This finding helps us qualify Bankoff's statement in *Cultures of Disaster*, where he says, 'Societies here have come to terms with hazard in such a way that disasters are not regarded as abnormal situations but as quite the reverse, as a constant feature of life' (Bankoff 2003, 53). In light of the data above, the everydayness and normalcy of 'disaster' pertains only to the experience of the Filipino poor. For the upper class, disasters are still experienced as disruptions to daily routines and intrusions into their zones of safety. 'Coming to terms with hazard' is indeed a psychological resource that the privileged upper class has not been forced to acquire.

Now that audiences' different consumption patterns and their significance to their experiences and interpretations of suffering have been established, the next section presents a case study of audience reception of a particular event of suffering.

Distant Suffering: Filipino Victimhood as a Moral Justification

This case study on audience responses to news about the May 2008 Sichuan earthquake attempts to provide a grounded perspective on philosophical arguments about people's moral obligations to distant others (Etzioni 1995; Singer 1973; Smith 1998). At the same time, it presents a non-Western perspective on Western-centric studies about the cultural favouritism of media producers (Chouliaraki 2006; Cohen 2001; Galtung and Ruge 1965; Moeller 1999; Orgad 2008) and audiences (Dalton et al. 2008; Höijer 2004; Kyriakidou 2005, 2008; Norgaard 2006). This case study uncovers that, unlike their reception of an entertainment programme, people's talk about suffering in the news features more explicit discourses of moral justifications. News, perceived by audiences as a more trustworthy genre than reality television or talk shows, provokes more rationalizations of action or non-action toward sufferers represented by media. I also reflect here that group interview dynamics functioned like a 'moral dilemma interview', where participants strategically deployed justifications in order to present socially desirable selves.

On 12 May 2008, a magnitude 8.0 earthquake hit the Sichuan province in China, claiming around 90,000 lives and destroying over 15 million homes. It was the deadliest earthquake to have hit China since the 1950s and caused additional deaths from aftershocks and landslides (Barboza 2008). The earthquake also came nine days after Cyclone Nargis in Burma, which caused 50,000 casualties. Media outlets such as the BBC ran comparative reports that documented government response and humanitarian aid in both countries: whereas in Burma 'people sat in the wreckage of their homes [...] often help was not on the way,' in China 'it has been a model of disaster relief' (Danahar 2008). The Chinese government's response was praised for its swiftness and openness, whilst the country in general was noted for its 'wealth' and 'resilience of infrastructure'.[2] The only significant criticism centred on the limited access to actual disaster zones granted to foreign journalists (Kendall 2008). Both disasters were covered in the world news segments of Philippine prime-time news using international news agency footage and Tagalog voiceovers.

In my sample of ABS-CBN's *TV Patrol World* episodes from 12 to 16 May 2008, I found a total of six separate news clips about the Sichuan earthquake in those five days. Five of these news clips were roughly 30 seconds long and were part of the 'World Patrol' segment: these news clips aired alongside other short clips of world news, such as Japanese technological inventions (14 May) and Angelina Jolie and Brad Pitt expecting twin babies (15 May). The five earthquake clips featured footage of ruined buildings and people stuck in rubble. Through the days, clips featured escalating numbers of casualties and used graphic language ('the leg of a child trapped under rubble required amputation'). Though no interviews with Chinese victims or officials were aired, there was footage and mention of government response and rescue operations. Whilst the presence of charitable actors conveyed some level of agency in the news clips, for the most part, they followed conventions of 'adventure news': sufferers were not given voice, they were depicted as mere aggregates, and the voiceover registered facts with no invitation for empathy (Chouliaraki 2006, 97–8). Overall, the news clips represented sufferers as 'too far' from audiences' sphere of responsibility (Silverstone 2007, 48).

Aside from these five news clips, there was one two-minute news clip on 14 May that discussed domestic concerns about the earthquake. The clip presented speculation about a strong earthquake hitting the Philippines, given that a magnitude 5.4 earthquake had hit the provinces of Isabela and Aurora in northern Philippines on 13 May. It discussed a viral text message about 'a US Geological Society' prediction of an earthquake, proven to be a hoax by a scientist in the news report. It also featured a short interview with a Filipino-Chinese woman expressing worry about the possibility of an earthquake hitting Manila. The clip ended with tips on 'What to Do during an Earthquake.' Analysing this using textual ethics categories, it is possible to say that this clip once again cast distant sufferers as 'too far' from audiences' horizon of action (Silverstone 2007, 48). Though this clip conveyed more emotion than the previous ones, it expressed anxiety and concern only for potential sufferers nearby rather than existing sufferers far away.

In the news-oriented group interviews that I conducted between June and August 2008, I used two 'World Patrol' clips and the '*Lindol*' ('Earthquake') news clip as materials to stimulate discussion and prompt recall of the event. After participants viewed the news clips, I invited them to share their views about the earthquake. My intention was to facilitate a spontaneous but focused discussion about natural disasters, poverty, the news and Filipinos' obligations to distant and nearby sufferers.

Across all focus groups, I learned that everyone had prior knowledge of the disaster. Lower-class audiences learned about the event through television, whilst the majority of the middle- and upper-class audiences learned about the earthquake through receipt of the viral text message, which prompted information-seeking through television news. Though they knew that many people had died, only those who consulted alternative sources of news (such as online news sites) were able to give good estimates of the number of casualties. This suggests that television news was unable to provide enough information and draw attention to this distant disaster, given its low priority in the newscast even during its immediate aftermath.

The clips nonetheless elicited respondents to express emotions. Across all groups, there were expressions of sadness ('I felt sorry for them'/'Sometimes I don't know why God allows these things to happen'). These tender-hearted discourses of compassion (Höijer 2004) were directed to aggregates rather than particular victims, and I suspect that this is a function of the 'adventure news' treatment of the disaster, which failed to give name and voice to the victims. But whilst sorrow was expressed for the Chinese people, my respondents across classes also articulated rational evaluations of the Chinese disaster, which involved assessment of the gravity of the disaster and evaluation of the authenticity of its victims in reference to the local context. Authenticity here though was not about sincerity, as it was in a playful genre in the last chapter, but fundamentally about the helplessness and victimhood of sufferers.

Lower- and middle-class audiences used material indicators, derived from popular images and stereotypes of China, to judge the earthquake victims and Chinese people in general.

LITO: Their buildings are sturdy and strong anyway. Compared to ours, which are dilapidated already.

KIKO: The materials that they use are superior for sure. You know, they can withstand the force of earthquakes. Don't they have the Great Wall [of China]?

(Group interview – Mandaluyong)

Even though the images from the newscast featured buildings reduced to rubble, they asserted that Chinese entrepreneurship and ingenuity would enable them to rebuild and recover quickly. These responses were informed by their knowledge of Chinese people in Philippine society, where they are known for their business acumen and high economic and political status (Pinches 1999). Their assumption was that, just like the Filipino-Chinese, the Chinese in Sichuan possessed enough material resources to help themselves.

However, just as they constructed Chinese sufferers as having more agency compared to sufferers in the local context, there was also compassionate acknowledgment of the difficulty that the earthquake victims faced. Here, they equated the way the Sichuan sufferers would experience difficulty with the way that 'the rich' in the Philippines would: 'Rich people like them have a harder time [coping] during times of crisis. Just like [rich people] here.' This response confused me at first, as it contained an identification of shared vulnerabilities between people 'like us' and 'not like us' – perhaps an indicator of cosmopolitan empathy. And so I probed this further:

> DORIS: It is a tough situation for them. But that's only at first. They'll find that life actually goes on and they can recover. I remember my boss at work – his warehouse burned down and so his family had to sell their cars and move their daughter to a [school with lower tuition fees]. I was so worried for them, that they might, you know, commit something bad because they haven't experienced that before. But I also want to tell them, hey, you should be thankful that your daughter is still in school and you're actually okay!

(Group interview – Mandaluyong)

I discovered that even though my respondents recognized common vulnerabilities between 'far' and 'near' 'rich sufferers', the evaluation remains that such suffering is still not as grave and authentic as poor people's suffering. The suffering of the rich is constructed as temporary and fleeting, because they assume that affluent people have resources to recover from setbacks. Though there is compassion here, it still constructs poor Filipinos' suffering as more deserving of symbolic attention and material reward. In the context of a moral dilemma interview, this construction can be interpreted as a deployment of a moral justification that allows the respondent escape from admissions of obligation or guilt (Seu 2003; Norgaard 2006).

Middle- and upper-class respondents also expressed that they were unable to directly aid the earthquake victims, although the justifications that they gave were slightly different from those of lower-class respondents. Both groups asserted their knowledge of China as a superpower in the world economy. Both cited the (then upcoming) August 2008 Olympic Games as exemplar of Chinese material resources, especially when it comes to infrastructure. However, some also displayed familiarity with the poverty and inequality that exist within China. In one upper-class group interview, there was debate about Chinese poverty in relation to Filipinos' obligation to help:

> MARIELLE: I'm sure they can recover. I have faith in them.

> JERIC: But I'm also slightly skeptical that, you know, China wants to brag that they're able to recover. But maybe they're really not

>helping their citizens that much. I mean, isn't that where they abuse sweatshop labourers?

MARIELLE: Well, yes. But with foreigners coming to China for the Olympics, won't it be obvious if [the government] leaves the destruction unmanaged? So surely, they're pressured to fix it!

JERIC: I hope so. I mean, all we can do is hope, really.

JONATHAN: Aside from hope, do you think you can also do anything else?

JERIC: You mean us? The Philippines? What can we send? Maybe our nurses or doctors. But even then, our provinces need our doctors more! We're too poor to help!

(Group interview – private university)

Compared to lower-class respondents, middle- and upper-class respondents have a more rounded and less homogenized picture of China. They were able to articulate conditions of suffering and poverty that were similarly authentic in terms of scale or gravity. However, such depth of understanding still did not translate to admissions of obligation to charitable action.

In the excerpt above, Jeric uses the Philippine government's own lack of response to deflect a question that attempted to probe his own lack of response. Once again, we can interpret this as a moral justification in the context of a group interview. Unlike the lower-class audiences who gave a singular but strong moral justification for their inability to donate (namely, their own poverty), middle- and upper-class respondents came to articulate more diverse rationalizations for inaction. Here, Jeric uses an attribution of inaction and helplessness to a larger collective (the Philippine government) to explain his own inaction. Other studies previously documented that such strategies are motivated by psychological needs of individuals to maintain a positive self-image (Norgaard 2006; Seu 2003). But in this context, we can also interpret this as a social strategy whereby upper and middle classes re-appropriate the poverty of other Filipinos to claim (temporary) ownership of victimhood in the context of a moral dilemma interview. Doing so allows them to provide moral justifications for inaction and to maintain a positive self-image not only to themselves but also to other participants and the researcher in the group interview. When middle- and upper-class members assert that they share in the same national condition of suffering, they strategically deflect discussions of obligation and painful admissions that they live on the right side of the divide between zones of safety and danger. Just as we saw in chapter three, where middle-class respondents re-branded *jologs* culture to assert positive value for themselves, temporary reclaiming poverty as part of their own condition is a strategic performance of moral justification that accrues positive value for themselves.

It should be noted that upper-class audiences also recounted experiences of donating money to victims of previous distant disasters, such as the 2004 Asian tsunami. Other respondents who previously worked overseas also recalled donating to foreign charities in company fundraisers. Joan, a 52-year-old bank executive and active member of a local charity, recalled that she had monthly deductions from her salary that went to a charity in Africa that her previous company in the US had supported.

When asked about their inaction in the context of the Sichuan earthquake, some upper-class respondents criticized the Philippine media as the reason for the general public apathy: they criticized the manner in which news about the earthquake was 'sandwiched' between 'silly' news items and lacked invitation for viewers to donate. Kyriakidou (2008) has cited that in the Greek context calls to action (i.e., flashing telephone numbers) in Greek news coverage of the Asian tsunami successfully triggered some of her respondents to donate. In this context, the local news representation of the Sichuan earthquake as 'too far' from audiences' sphere of responsibility was reproduced in audiences' lack of attention, confirming Chouliaraki's (2006, 106) arguments about adventure news. And that audiences criticize the 'bad' quality of media representations of suffering in order to justify their inaction towards sufferers is a finding that resonates too with Dalton et al.'s (2008) research in New Zealand.

I found that middle-class respondents actually asserted *more* justifications of inaction than ones in the upper class. They remarked on the 'hypocrisy' of charity for distant others. For instance, call centre agent Benjie said, 'Shouldn't we try and clean up our own mess before we intervene with others? What are these people [who are more concerned about international issues] trying to prove to *us*?'

Practices of giving to strangers before the family or to non-Filipinos before fellow Filipinos were recounted as undesirable and problematic. These middle-class judgments resonate with McKay's (2009) own ethnography, in which overseas Filipino workers who ignored their obligations to their kin in favour of forging new ties in the host country were judged negatively. Additionally, my middle-class respondents gave examples of 'hypocritical' practices on social networking sites such as Facebook. They cited how some of their upper-class Facebook friends repeatedly posted status updates about foreign politics, distant disasters and foreign celebrity gossip instead of discussing local news. They regarded such practices as 'posturing' and 'social climbing'. In the context of talk about distant suffering then, communitarian discourses of mutual aid and cooperation among Filipinos are strategically used to assert middle-class moralities of authenticity tied to principles of nationalism and patriotism. Cosmopolitan charity is constructed as an upper-class strategy to assert high status and superiority over other Filipinos and is perceived as a form of 'cutting the network' (McKay 2009; Strathern 1996), whereby natural and immediate obligations to neighbour, community and homeland are cut in favour of forging relationships with distant others. Whilst I acknowledge that instrumental motivations behind cosmopolitan charity may apply for some upper-class donors, I argue that this generalization is nonetheless used as a moral justification by middle-class respondents in the context of the interview. By assigning a negative value to cosmopolitan charity, my middle-class respondents escape the painful admission that they may not be as good and generous as they would like to think themselves to be by de-valuing practices of their own proximal others – the upper class.

One form of action articulated across social classes was prayer. Among my Christian and Catholic respondents, prayer was constructed as a legitimate form of action in the context of distant suffering. Prayer here involves recognition of their shared humanity with distant others by virtue of both being God's creation, whilst at the same time acknowledging their difference. Praying for suffering others involves expressions of tender-hearted compassion ('God, please lift their burden'/'I pray that they turn to You and draw strength'). Following Catholic belief that Christ's image is most evident in encounters with the miserable (O'Brien 1993), it is not surprising that encountering distant sufferers in the news involves reflection about one's own relationship with God and an articulation of gratitude for the present moment.

Having teased out Filipino audiences' moral justifications in the context of a particular case study of distant suffering, I explore in the next section the conditions for charitable action and volunteerism towards local suffering in the news. This section focuses its discussion on the interaction between media's representational strategies as well as audiences' personal and social contexts as mutually informing their actions toward televised sufferers.

Complex Interactions: Textual Ethics and Audiences' Contexts

Whilst we have seen that moral claims are offered up by diverse representations of suffering in various ways, the ethnographic approach I adopt here enables us to identify patterns in the textual qualities of suffering that resonate with particular groups of people. Though Tester (2001, 71) has said that 'it is impossible to predict in advance whether, how, or which journalistic production will be or become morally compelling' for the audience, I argue that this study has been able to map out how audiences' class backgrounds orient them to particular understandings of suffering (lay moralities of authenticity, respectability and deservedness) and media conduct (lay media moralities). Using two case studies of audiences' charitable practices, I now highlight how these classed moralities of suffering and classed lay media moralities similarly inform, but do *not* determine, charitable action toward sufferers on television.

Charity appeals and the retail fundraiser

As mentioned, the institutionalization of the synergy between news and charity is peculiar to the Philippine media landscape. Whilst pundits generally regard this in positive terms (De Quiros 2009a), I found that my upper-class respondents are ambivalent, if not highly critical, about the link between charity and news. Though they acknowledge that provision of social services to the poor is generally desirable in light of weak government structures and the absence of a welfare state support system, they criticize the manner in which social services are dispensed by media charities. Similar with entertainment, the conditions of media pilgrimages in the context of news charities are evaluated as problematic. First, media pilgrims are evaluated as lacking in resourcefulness on the one hand and respectability on the other, just as they were regarded in *Wowowee*. Some upper- and middle-class respondents doubted poor people's willingness to work hard when they chose to accept dole-outs from rich media patrons and believed them to be complicit with what are perceived as exploitative media practices.

This brings me to the second issue: the rules that the media set to govern these media pilgrimages contradict upper-class lay media moralities and proper charitable conduct. They express disgust for media conventions that 'force' people to display emotions and make direct appeals to viewers – an attitude similar to what we have seen in chapter three of upper-class taste judgments of excess (see also Bourdieu 1986; Skeggs 2004). They perceive charity appeal conventions that have victims and/or their families speaking for themselves to request for viewer donations ('I call on viewers to have pity on my child to donate [...]') as an unnecessary imposition by producers on people already in obvious distress. The producers' motivations are questioned here, given that audiences are aware of the resources that television networks keep to themselves: 'Why don't the media themselves give money to the needy? Why do they need to request assistance from viewers?' asked Jordan, a university student.

Upper-class audiences speculate then that 'sick children stories' – because of their high emotional intensity – attract attention and score good ratings. It is interesting that 'giving a voice' and 'giving a name' to sufferers, in this context, does *not* enable compassion as previously speculated in the literature (Chouliaraki 2006; Cottle 2006). Upper-class respondents instead prefer that experts and authorities speak for the sufferers as well as manage their graphic visibility. The selection of beneficiaries certainly plays a role here: given that charity appeals tend to select children with illnesses and physical deformities (hydrocephalus is a disease perceived to be 'overrepresented' in media charity appeals), I have observed university students literally switch off the television during my visits in their homes.

I also found that upper-class donors or volunteers and middle-class employees of faith-based charity organizations such as *Gawad Kalinga* (Bestowing Care) were *more* critical of media charities than other upper- and middle-class respondents. The most religious among these respondents argue that the practice of charity by media is inconsistent with Catholic social teachings that emphasize the preservation and flourishing of human dignity alongside compassionate practices of charity. Some perceive that the God-given dignity of sufferers is violated when they are made to cry and beg on television by producers, reflecting once again their judgment on asymmetrical relations of exploitation between media and sufferer. Others criticize the perceived self-aggrandizing nature of media charities that flaunt their generosity on television. They cite verses from the New Testament ('But when you give to the needy, do not let your left hand know what your right hand is doing', Matthew 6:3) to emphasize that charity should be performed in private rather than in public. Some members compare television charity appeals with *Gawad Kalinga*'s own 'more ethical' communications materials. *Gawad Kalinga* charity appeals, used during fundraising events such as dinners and concerts attended by wealthy donors and guests, use 'empowered' and 'positive' images of sufferers (see Chouliaraki 2010, 112–14). Their videos commonly feature before-and-after images of their beneficiaries, showing the process by which slum communities have been transformed by the generosity of donors who pledged small concrete houses for individual families. Certainly, these videos portray higher levels of agency than charity appeals on Philippine television – agency that upper- and some middle-class *Gawad Kalinga* members desire to see in sufferers.

In contrast, lower- and middle-class respondents (among the latter, especially those who are not members of charitable organizations) tend to be more receptive to charity appeals. Whilst there were a few cases of middle-class respondents who claimed to have sent cheques and donated equipment to aid particular beneficiaries, the overwhelming response is that lower- and middle-class audiences recalled leaving coin donations in coin banks of ABS-CBN's *Bantay Bata* (Child watch). These coin banks of ABS-CBN's children-focused charity are located in bus and train stations, fast-food outlets and grocery stores. Some respondents mentioned that their donations were sometimes motivated by particular charity appeals that they remembered, but some also claimed that coin donations had become a habit and were therefore not motivated by specific appeals. Some mentioned more spiritual dimensions to giving: coin donations were their way of 'thanksgiving' for a comparably better life than what they see on television or in everyday life. Media networks, specifically their news and charity organizations, are highly trusted by low-income respondents. Given that my respondents have seen visual evidence of help bestowed to beneficiaries on television, they trust media charities with their donations more than non-media charities and NGOs. This high level of trust among my middle- and lower-class respondents toward media charities is in fact consistent with the findings of a 2003 Nationwide Survey of Giving, which ranked ABS-CBN's *Bantay Bata* as the most trusted charity (Venture for Fundraising 2003).

As we have seen in the previous chapter, low-income respondents evaluate media pilgrimages – and pilgrims – positively. They interpret respectability and resourcefulness in pilgrims' willingness to play the rules of the game in the media centre and attest to their difficult life conditions in public. They also evaluate charity appeals, in showing 'graphic' images of suffering, as *authentic* representations of suffering. By showing suffering 'at its worst' (Cohen 2001, 183; Orgad 2008, 21), charity appeals provide visual evidence of the depravity of the victims' conditions and show that they deserve to be on television over multiple suffering others. Media pilgrims, whilst acknowledged to be placed in a difficult situation of speaking about their conditions of illness and duress, are nonetheless perceived as having resourcefulness to find creative ways of saving the lives of their loved ones. As Joshua, a thirty-six-year-old carpenter in Park 7, told me,

> It is no joke to go and ask strangers and beg for money. When people go to the media, that means that's it, they're at knife's edge, they're ready to do anything. But actually admitting that to yourself and bringing your child with you to ask for money from the media [...] To me, that's bravery.

When I interviewed media charity representatives during fieldwork, I discovered that much of their income actually comes from coin bank donations, not from large-sum donations from private individuals or companies. Tina Monzon-Palma, ABS-CBN news anchor and head of ABS-CBN's *Bantay Bata*, disclosed to me that 80 per cent of the annual income of their organization comes from coin banks. She calls herself a 'retail fundraiser' in acknowledging that lower-class charity is the main driver of her organization's success. The implication is that charitable action in this context is not a mediation of capital that crosses upper-class zones of safety with lower-class zones of danger; media charity appears to be a loop of mutual aid and cooperation within zones of danger. This reflects traditional models of *bayanihan* that are practised within poor communities to ward off anticipated disaster (Bankoff 2003, 168–9). Additionally, this is consistent with our previous discussions of upper-class 'switching off'. Because the upper-class audience has a general orientation of disengagement from the public world of local television, they often do not receive invitations for charitable action through charity appeals. And because the upper class generally traverses the metropolis within zones of safety where these ABS-CBN coin banks are not often present, charity for them more often occurs outside the media, except during exceptional moments like Ondoy.

Media as government during Typhoon Ondoy

In chapter one, we opened with the assertion of media acting as the government during the tragedy of Typhoon Ondoy in September 2009 (De Quiros 2009a). For a culture of disaster, where catastrophe is supposedly an everyday occurrence, Typhoon Ondoy was greeted with a rare government declaration of a 'State of Calamity' in the nation's capital. In the words of some journalists, Typhoon Ondoy was the 'great equalizer' in submerging 80 per cent of the city under water, zones of safety united in tragedy with zones of danger:

> The flood treated everybody, rich and poor, equally. It didn't play favourites, exempting no one. It made everybody miserable. Celebrities and the influential suffered along with the poor. The rich in their gated communities suffered just as much as the squatters in their shanties. The relatively well-off Provident Village in Marikina was among the worst hit, with floodwater reaching the rooftops (Cruz 2009).

In a review of television coverage of the Typhoon Ondoy tragedy, the first news clips contained images of heavy flooding, cars stranded and men wading through chest-deep waters. Increasingly, the unprecedented nature of the disaster was conveyed through live interviews with celebrities, panic-stricken on the telephone and crying for rescue as floodwaters rose: 'Rescue, please! The water's about to reach our roof! We don't have light or medicine!' cried teen actress Cristine Reyes – whilst a still photograph of her frail figure crouching on the rooftop was flashed. Television news also picked up on user-generated videos on YouTube of massive flooding in posh neighbourhoods as well as Facebook status updates of Manila's elites condoling with victims and expressing shock – a few calling for assistance for their own family and friends.

The day after the typhoon, both networks launched telethons interspersed throughout the day's programming. Heavy criticism was directed at the government for their lack of disaster preparedness: the networks pounced on the story that the Philippine Navy owned fewer than thirty rubber boats, impeding search and rescue operations on still-flooded streets of the metropolis. Talk show host Ruffa Gutierrez spoke while co-chairing the telethon: 'You government officials, I wish that you would just stop talking. Go out and help. Other people are helping already. [Celebrities] are rescuing people. My assistant is rescuing people. Please, it's time for you to do something.' ABS-CBN quickly announced that *Wowowee*'s Willie Revillame had lent his personal helicopter to the government for rescue operations. Media personalities took it upon themselves to call out corporate sponsors and request donations from them, even shaming companies that had not yet pledged donations on live television (e.g., 'Avon, how come you haven't donated? Other brands have donated.'). News anchors and charity heads opened their warehouses to invite the public to volunteer as packers or distributors of relief goods or as telethon phone operators, and streaming images of volunteers in media centres were juxtaposed with images from the field and talking heads in the studio. In less than a day, ABS-CBN boasted cash and kind donations amounting to 30 million pesos (430,000 pounds), whilst GMA raised around half this amount.

The relief efforts of media charities eclipsed the efforts of other institutions, such as private universities, the Philippine Red Cross (PRC) and even the government. Whilst short clips of the relief efforts of government agencies and the Malacañang Palace (headquarters of the Philippine President) were aired, these contrast to the scale and spectacle of donating and volunteering within the media centre. News clips about failed government rescue operations in flooded areas appeared in stark contrast with the charitable 'reverse pilgrimages' of ABS-CBN and GMA journalists and celebrities to disaster zones. These 'reverse pilgrimages' typically featured large and happy crowds queuing for relief goods being handed out by journalists and celebrities.

Days after Ondoy, I tried to visit several low-income respondents in their communities. In squatters' areas that I visited, many families had moved out of their houses and brought their recovered belongings to basketball courts and public schools in their areas. These public spaces became refuge centres. In *Bagong Silangan* (New Horizon), one district that I visited in Quezon City, half of the basketball court was lined with the coffins of those who died from the floods; the other half was where the homeless lay in their cots. The mood in these areas was not at all bleak: groups of children played hide and seek, people sat in groups sharing stories, and crowds excitedly gathered around all incoming private vehicles in anticipation of politicians, media charities, celebrities and other concerned donors bringing food, clothing or cash. Several low-income respondents cautioned me about potential stampedes: the mad rush to queue for free lunches and loot bags had injured several people already. The scene in these public spaces was definitely more frenzied – even unusually festive – compared to my attempted visit to the houses of several informants. Walking through streets ankle-high with mud, I observed residents silent and exhausted whilst cleaning and scrubbing the walls of their homes and what was left of their personal possessions: teddy bears, photographs and mattresses laid on pavements, partly obscured by dried mud. Talking to some of my respondents, I was nevertheless surprised to hear them make hopeful predictions about when their homes would be clean and ready to be lived in again. Unlike some of my middle- and upper-class respondents, not one of them was tearful and overtly sentimental about the loss of personal possessions; after all, they personally knew neighbours who had lost family members in the floods. Some of my respondents' children even expressed surprise when I told them that the private university where I worked cancelled the remaining weeks of term: 'Really? But my friends and I are about to begin working on our homework! I wish our classes were canceled too!'

Given that electricity in many squatters' areas was still irregular, my lower-class respondents' experiences with television were more characterized by direct interactions with media charities, celebrities and journalists. Once again, they regarded the services (in this case: free food and/or gift bags of clothes and personal care items) that they provided as greatly helpful. The only negative statements from some low-income respondents were about the limited reach of the distribution of these material goods. Those who were cleaning their homes rather than waiting in the public areas could not queue up and access these goods and services. Some of my low-income respondents who lived in harder-hit or less populous areas complained too that the media 'never visited' them. Certainly, some squatters' areas were more difficult to access by large live-broadcast vans because of their narrow (or muddied) streets or were less visually interesting (e.g., they lacked a large public space where journalists could do live reports) and thus did not experience the services offered by the television networks. This feeling of exclusion or abandonment from the media's charitable efforts did not affect their overall judgment of the media, however. They still acknowledged – and trusted – that the media were engaged in judiciously evaluating which areas were hardest hit by the typhoon and therefore most deserving of help.

As for my upper- and middle-class respondents, I found that some, but not most, were directly affected by the floods. However, all of them knew of friends and relatives directly affected by the tragedy, their entire homes submerged underwater.

For upper- and middle-class respondents not directly affected by the floods, the shock of the tragedy did not actually come from television images of the event, but from words and photographs exchanged on Facebook, YouTube and their mobile phones. Though their personal experiences of discomfort during the typhoon (e.g., being stranded, losing electricity) alerted them to the possible magnitude of the disaster, the assumption was that, as was typical, the disaster would have no significant impact on people they knew. It was only when they saw rich friends posting photographs of their flooded houses and texting or posting calls for help on Facebook that they realized the unusual nature of the disaster, thus provoking a heightened emotional response. As university student Julia said, 'When I saw my friend's Mercedes Benz underwater, that's when I knew this was something different! If she's affected, then I'm sure almost everyone is!' A few also recounted some trauma in anxiously trying to contact and search for friends living in worst-hit areas, unable to respond to media communications days after the typhoon.

For my private university respondents, their university canceled the remaining three weeks of term in sympathy with students and staff directly affected by floods as well as to encourage student and staff volunteering in the university's relief efforts. All of my private university respondents shared experiences of helping friends affected by the disaster, volunteering in the university or visiting ABS-CBN and GMA Network. My older respondents also described helping friends and donating to either media charities or charities that they had had prior affiliations with.

Interestingly, my upper-class respondents all evaluated the media charities as being 'efficient' and 'trustworthy' in the specific context of Ondoy. Many cited their speed of fundraising and distribution of relief goods, from the images that they saw on television. They mentioned too the ease in donating to media charities, given that telethon hosts repeatedly provided information about the process of donating and volunteering. Upper- and middle-respondents resonated too with repeated messages of unity, togetherness and compassion that ABS-CBN and GMA hosts circulated during the tragedy (De Quiros 2009b), more so than with the Philippine President, who gave no public speeches and only granted brief interviews to journalists.

The material resources of television networks, combined with the media's unique symbolic power, indeed constructed the media centre as the most valuable and desirable space in which suffering was resolved and personal anxiety managed during this unprecedented crisis. Government efforts could not but be unfavourably compared by audiences to those of the media institutions. First, media institutions circulated anti-government rhetoric that has become a traditional trope in local journalism. Second, the media offered alternative social services of their own. Media's role as mediator here entailed

Figure 2: Students and alumni of Ateneo de Manila gather in their sports centre to pack food and other relief goods for affected communities of Typhoon Ondoy.

Photo by the author.

not only mediation of narrative frames, but economic mediation as well: they invited, received, converted and redistributed capital from individual donors as well as the private companies that were their advertisers and sponsors. Third, television attempted to provide 'ontological security' (Giddens 1991;

Silversone 1994, 2005) through their formal structures of narrative closure and scheduling and also through recirculating, even amplifying, the standard therapeutic discourses of togetherness and resilience in conditions of great suffering that factual entertainment programmes espouse.

Rather than 'switching off', upper-class respondents could not help but tune in to local television during this crisis. This was motivated in part by the perceived absence of functional government structures and the low visibility of other civic organizations in society. However, it must be noted that the upper-class tuning in to local television could not have been possible without the array of new communications technologies and platforms – mobile phone, Facebook, YouTube – that personalized the event of suffering and gave face to the sufferers, some of whom were part of their social networks. Lay narratives of suffering from their friends on social networking sites drew their attention, provoked emotional responses of compassion and prompted charitable action. Upper-class positive evaluations of the media here were concerned more with the processes of donating and volunteering within the media centre than their actual representations of suffering in the news.

News: Negotiating Moral Claims

Unlike the entertainment programme discussed in the previous chapter, the news genre is more diverse in its qualities of representing suffering. Though it similarly follows practices of 'overrepresentation' (Wood and Skeggs 2009, 177) consistent with other genres, the news portrays suffering nearby and distant, personal and collective, using varying emotional registers and deploying different material resources. As a genre cross-culturally more trusted than others (Livingstone 2005; Madianou 2005b) and enjoying high levels of respectability and authority in a country with an esteemed tradition of journalism, the news is more provocative than entertainment when it comes to audiences' working through compassion and charitable action for televised sufferers. At the same time, its formal structures may provide people 'ontological security' (Giddens 1991; Silverstone 1994, 2005), especially those who identify as suffering themselves, to manage their personal anxieties in the face of crisis and catastrophe.

The generic structures, local industry conventions and cultural significance of the news interact once again with audiences' contexts, particularly their class-informed moralities of suffering and lay media moralities, in shaping audience responses to news texts. We see more occasions of tuning in from the upper class when it comes to the news and more varied discourses of compassion among audiences of different classes. What is consistent is lower-class audiences' continued dependence on and trust in television for providing them symbolic and material resources for coping.

Table 3. Mapping Moral Responses to Suffering in the News

Audiences	Consumption of News	Discourse of Compassion	Judgment on Sufferer	Judgment on Genre-Sufferer Relationship	Criteria of Deservedness
Upper class	Occasional viewing (motivated by information-seeking of events of suffering) (younger people more likely to be shocked and disgusted by news)	*usually blame-filled compassion (directed at perpetrators; can also be directed at media that publicize or address suffering)	Varies (distinction between nearby and distant sufferers; also minimal distinctions among nearby sufferers)	*news as legitimate space to make visible suffering; though too frequent and graphics images of suffering may prompt 'switching off' *news as legitimate space for donating and volunteering during public crises in the absence of functional public institutions *members of religious charities such as *Gawad Kalinga* more likely to criticize media charities	* nearby sufferers more deserving than distant sufferers * preference for empowered sufferers and positive images * attention driven by ecstatic news
Middle class	occasional to regular viewing	* tender-hearted and blame-filled compassion	Varies (distinction between nearby and distant sufferers; also distinctions among nearby sufferers)	*news as legitimate space to make suffering visible, especially when it pertains to local conditions or problems that can be avoided or managed	*nearby sufferers more deserving than distant sufferers *donates to media charities *tends to be critical of 'cosmopolitan' charity

| Lower class | Affective consumption (news regarded as documenting everyday conditions of sufferers) | *tender-hearted compassion (directed to individual sufferers) | Varies (distinction between nearby and distant sufferers; also fine distinctions among nearby sufferers) | *news as legitimate space to make suffering visible, especially when it pertains to local conditions or problems that can be avoided or managed
*news as legitimate space for donating and volunteering in the absence of functional public institutions | *news as legitimate space for donating and volunteering in the absence of functional public institutions
*members of religious charities such as *Gawad Kalinga* more likely to criticize media charities | *responds to pity-based claims of suffering by donating to media charities |

This bottom-up investigation into televised suffering attests to the significance of textual tthics approaches to the news. Editorial decisions, such as whether to accord more or less visibility to certain events, or whether to use emotional or purely factual narration of suffering indeed influence attention and compassion, as we saw here in Filipino responses to the Sichuan earthquake. Whilst not an ideal type of 'adventure news' in Chouliaraki's (2006) standards, as it in some way depicted some degree of agency in the efficient Chinese response to the disaster, local news coverage of this event of distant suffering disabled perspective taking, information seeking and working through of decisions to act. Especially when situated in a cultural context with established traditions of mutual aid, cooperation and *bayanihan* (Bankoff 2003) and sanctioning of decisions to prioritize cosmopolitan connections over local kinship ties (McKay 2009), the news reinforces rather than challenges the communitarian status quo. This lack of cosmopolitan imagination in Philippine news does little to eradicate stereotypes about distant others, which we saw many lower-class respondents reproduce in the interviews above, and according to Cabañes' (2013) ethnography, it may inform and legitimize everyday practices of racism towards foreign tourists, cultural minorities and migrants in Manila. Textual ethics discussions about mediated suffering, though understandably Western-centric in the assumption that privileged Western publics have more resources to aid non-Western sufferers, are similarly needed in non-Western contexts such as the Philippines, where local audiences are also involved in the ethical question of how to engage with near and far others.

A key theme that emerged in this chapter is the working through of decisions to act on suffering others – that is, acting beyond 'effective speech' (Boltanski 1999, 18–19) and actually donating or volunteering in response to media representations of suffering. The default upper-class media orientation of 'switching off', we find here, prevents any form of decision-making as to whether to act or not toward everyday news items of poverty. This closure from the public world mediated by Filipino television, along with their lay media moralities that media pilgrimages involve an aspect of exploitation of individual sufferer by wealthy corporation, disables them from making modest contributions to media charities. This is not to say that upper-class respondents do not practise charity. After all, we saw that *Gawad Kalinga* members practise charity outside the media, where their Catholic beliefs about human dignity, contracted through classed standards of privacy and respectability, are in resonance with the representations and resolutions of suffering by the organization.

In light of upper-class 'switching off', media charities have become retail fundraisers that recirculate capital among the middle and lower classes, rather than involve cross-class transaction. ABS-CBN's charity *Bantay Bata* draws 80 per cent of its yearly income from small-denomination donations in coin banks present in many middle- and lower-class zones in the country.

It is difficult to make a judgment on whether upper-class lay media moralities of exploitation are definitively correct or not, especially when situated side-by-side with lower-class lay media moralities that assert that gruesome representations of suffering are actually authentic and truthful. Also it is interesting to judge whose practice of charity, if any, is correct among the upper and lower class here: whilst most respondents share the same religion of Catholicism – one whose 'great virtue' is that of charity (Comte-Sponville 2001, 289) – their translation of charity from belief to practice is greatly different and markedly classed. Though this study, inspired by the anthropology of moralities, does not aim to pass judgment about these contending moralities, it could suggest at a practical level that media charities need to develop alternative ways of representing of suffering if they intend to invite greater quantities of upper-class charity.

Even though Ondoy became a context of cross-class charity mediated, even orchestrated, by television, it was also an exceptional tragedy characterized by an 'ecstatic' and communitarian discourse that 'we are all victims' (see Chouliaraki 2006, 178–81). Upper-class compassion was in fact initially provoked by personal knowledge of people directly affected by the disaster, facilitated by new media, that only later led to tuning in to the television and evaluating it as a desirable space for donating or volunteering.

Whilst Ondoy was indeed exceptional in its intrusion into traditionally fortified zones of safety in urban areas of the Philippines, as an event of suffering it was actually quite ordinary for the poor. For the poor, everyday suffering is about having little, being uncertain about how to make a living and where to find the next meal and hoping that the compassion of others may make their lives just a little easier. Though scholars, government officials and the media themselves have long attested to the resilience and coping mechanisms of the Filipino poor, the terms 'resilience' and 'coping mechanisms' ultimately just describe what the poor actually do to survive. As De Quiros (2009b) says,

> Because except for certain calamities like Ondoy, which ravage both rich and poor, typhoons and floods tend to ravage largely the poor: the ones protected only by the frailest of roofs from the howling winds; the ones who live in the path of mudslides and [volcanic debris], having

nowhere else to go; the ones who huddle in along creeks and [sewers], where they've strayed after having been driven away from every available empty space beside government buildings.

Just as governments in democratic societies require engaged citizens, the media depend on engaged audiences across classes if more effective, more democratic and less unequal political, economic, social and cultural institutions are desired. Only a public orientation that involves attention, interest and empathy for others could provoke audiences to begin thinking about and acting on televised suffering, whether exceptional or everyday.

CONCLUSIONS: MEDIATING SUFFERING, DIVIDING CLASS

The test of acknowledgment is not the reflex reaction to a TV news item, a beggar on the street, or an Amnesty advertisement, but how we live in between such moments.

Stan Cohen, *States of Denial*

This book has explored moral problems in the mediation of suffering using an audience-centred approach. Drawing on an ethnographic study of television audiences in the Philippines, this book provided empirical material in dialogue with the normative and often text-centred media ethics literature. Inspired by perspectives within the anthropology of moralities, it sought to be attentive to moral discourses and practices in the context of everyday life whilst keeping these in conversation with normative theory as well as empirical studies of other cultural contexts. At the same time, it has demanded more careful evaluation of television production of suffering – one that acknowledges the diversity of media institutional practices around the world and differences in audiences' social positions.

As the title clearly made the point, one of the central themes of this book is the discussion of the poverty of television in a class-divided Philippines, anchored from a holistic yet insistently grounded account of audiences' (dis/) engagement with media and their varied ways of over-representing and even resolving poverty and disaster. This sheds light on particular challenges of the Philippines as a developing country where, social denial strategies notwithstanding, it is most impossible for people to honestly claim ignorance about the plight of vulnerable others (Bauman 2001, 1); after all, suffering exists in excess both on national television and in the 'liquid mess' of the metropolis (Tadiar 1996). In this context, the poverty *of* television thus is not an 'out there' phenomenon ('distant suffering' in the Western-centric literature) but 'in here' – an ever-present social reality that television culture shapes and is fundamentally shaped by. The everyday experience of poverty by the lower-class majority is crucial to the political economy of mass media institutions, their motivations for legitimation, their programming styles and stories, and the ways in which they invite media pilgrimages from charity-seeking poor people.

Television's centrality in Filipino public life is seen to go beyond its symbolic power to describe and represent reality (Thompson, 1995), but also in the media's very interventions to solve and address suffering in the place of a weak state. The institutionalization of charity into television entertainment and news production is employed as branding strategies to appeal to viewers and loyalty strategies to invite audiences into an extended television family network with obligations of eternal viewer loyalty. This book uses this messy and complex backdrop of unusual and unregulated media practice to reflect on the tenability of media norms and values in diverse social situations rather than dismiss television as a failed institution for 'moral education' (Chouliaraki, 2008). Television has been analysed in this book in its diverse articulations: on one hand as communications technology with multi-ranging affordances, and on the other hand as institution that reorders, regulates and redistributes attention and care.

Given the local context of the liquid mess of everyday suffering amidst rigid social stratification, it is significant to develop an analytics of class in the production and reception of suffering. Class-divergent television consumption practices necessitated explanation of poor people's strategic suffering on television (see chapter four) and elites' disengagement from their undesirable others (see chapter three). The book likewise observed that class influences television consumption as much as television shapes and reproduces class: in Philippine television's practices of over-representing the poor and under-representing the elites, the media maintain and at times amplify class divides.

This study arrived at these observations and conclusions through the tools of ethnography, exemplifying a more holistic understanding of (1) the social process of mediation and (2) the actually existing scenarios where abstract philosophical norms and normative positions of good and bad media practice are enacted.

On the first point, the ethnographic and grounded nature of this inquiry provided a more nuanced description of the process of mediating suffering than in earlier speculative accounts. For instance, postmodernists have assumed that television produces spectacles and simulations of human suffering that lead to audience enchantment and moral numbing (Baudrillard 1994, 2001; de Zengotita 2005). Within the media ethics literature, there have also been arguments about how 'compassion fatigue' is an 'unavoidable consequence of the way news is now covered' (Moeller 1999, 2), but these come from reviewing journalists' practices without taking into account the voices of audiences. The ethnographic approach taken here revealed that audiences' experiences with televised suffering are characterized more by emotional expression and moral reflection than by 'moral numbing' and 'compassion fatigue'. And whilst some people did avoid televised suffering, this practice was shown to be motivated by the audiences' class positions rather than purely determined by television as technology or the news as genre. This study affirms Livingstone's (2009) and Madianou's (2009) position that ethnography is the methodology best suited to investigate in-depth the multifaceted nature of the mediation process (for further discussion of the study's methodology, see the Appendix).

On the second point, this ethnographic study uncovered moral positions that people take in relation to media and suffering rather than merely recording the moral position of the researcher. This is important for four reasons.

First, it allowed comparison and contrast with Western theorists and moral philosophers' position that audiences are responsible for and should care for distant suffering others (Silverstone 2007; Singer 1973, 2009). We have found that television talk of non-Western audiences in the Filipino case did construct televised suffering as a moral dilemma, though in some cases it did not do so. Distant suffering has been the focus of Western media ethics, but in the Philippines, proximal suffering presented the most urgent dilemmas, those that evoked the strongest emotions and moral judgments. The comparison and contrast between moral issues of distant versus proximal suffering was enabled by audience-centred ethnography and affirmed the significance of an anthropology of moralities that identifies local moral codes important in people's lives instead of merely imposing the external codes of the researcher (Heintz 2009).

Second, an ethnographic approach allowed for 'following the trail' and 'surprises' (Strathern 1999, 9) that led to the discovery of salient contexts of moral reflection about mediated suffering. The focus of conventional research on the news genre has occluded moral issues in the production and reception of suffering in other genres such as factual entertainment (found here to be most provocative of emotional responses to others' suffering) and local industry practices of addressing suffering within the mediated centre (found here to inform audiences' overall evaluation of media ethics). Ethnographic analysis enabled us to make connections as to how different media texts, genres and practices collectively and individually invite moral judgments from audiences.

Third, as elaborated in detail further below, an ethnographic approach contributed new sets of questions that can be used in the ethical critique of mediated suffering. For instance, it led to the empirical finding that the representation of the agency of sufferers on television does not determine a singular response from audiences, as the relationship between media texts and audiences' contexts is much more complex and dialectical. Ethnography could then provide more expansive consideration of other moral consequences of media texts and practices, such as the widening of social divides within one nation, and could suggest new questions for theorists to ask when proposing normative formulae to ethical inquiry of media and their audiences.

Fourth, the bottom-up nature of this book recorded the voices of marginalized groups – *masa* television audiences frequently described, derided and infantilized by academics, critics and even the media themselves (Contreras 2011; Dancel 2010; De Jesus 2011; Tan 2006). The low-income participants of this study represent too the sufferers in non-Western zones of danger, who are often referred to but rarely heard in the media ethics literature. In scholarship that aims to shed light on issues affecting marginalized groups, the recording of victims' voices may 'speak more meaningfully' about the experience of pain, according to Das (1995, 176). As existing public discourse and academic scholarship privilege expert opinion and moralities, an 'anthropology of pain' that 'provid[es] voice' and 'touch[es] victims' 'so that their pain may be experienced [by others]' is also necessary (196). It is my hope that the recording of marginal voices in previous chapters contributes to challenging stereotypes, inviting dialogue and enabling discourse about social suffering that is hospitable to those who suffer themselves.

I now summarize my key findings and contributions to different sets of literature.

Everyday Suffering

Most of the participants of this study would have belonged to the category of distant others in the Western-centric literature on media ethics and distant suffering. They would likely be referred to or perceived as beneficiaries of (Western) charity, and scholarship would have primarily asked how global television could bestow greater visibility to their cause. However, the empirical focus on the Philippines entailed a reversal of angles and considered how non-Western audiences themselves make sense of televised suffering. This move has allowed for productive dialogue with the literature by discovering similarities *and* differences between Western and non-Western experiences in the spectatorship of suffering.

Filipino upper-class respondents who avoid local television were found to share much in common with some of Kyriakidou's (2005; 2008) middle-class Greek audiences and Seu's (2003) British respondents who routinely turned away from images of suffering. These privileged audiences negotiated similar (though not the same) moral dilemmas of acting or justifying their lack of action toward televised suffering. Avoidance strategies and moral justifications are present cross-culturally: though Western audiences 'switch off' because suffering is too strange or too far, privileged Filipino audiences 'switch off' because sufferers are too threatening, too many and too near.

However, the majority of my respondents who lived in poverty and identified themselves as suffering provided an interesting contrast to existing scholarship. My low-income respondents' affective consumption of television would challenge conventional assumptions that greater and repeated exposure to images of suffering would lead to 'desensitization', 'indifference' and 'compassion fatigue' (see Cohen 2001, 188–91). We found that lower-class Filipinos do not avoid suffering in the news and entertainment – rather, they seek it out. They do not use television to escape to fantasy worlds of opulence but to actively seek compassionate practices of recognition and redistribution for sufferers like them. They consider television's practices of over-representation of suffering as symbolically and materially beneficial but hope that only the authentic and deserving sufferers are given visibility and granted reward. This was evident in chapters four and five when the poor expressed judgments of authenticity and deservedness toward sufferers on television.

Suffering for the poor is not so much about 'famine, death, war, and pestilence' as it is in Western descriptions (Moeller 1999, 1) and Filipino upper-class experience, but it is an everyday condition – their personal and collective property, experienced in varying degrees. In this context of everyday suffering, I argue that audience responses to images of suffering include both experiences of avoidance/shock/indifference and affective consumption involving identification, hope and conferment of moral judgments. This everyday experience of poverty, as well as historical, social, cultural and religious beliefs about suffering, shape audience responses, just as the textual ethics of television representations influence responses.

My interviews with lower-class people reflecting about everyday suffering share parallels with Melhuus' (1997) ethnography of interpretations of suffering in Christian Mexico as well as Abu-Lughod's (2002) study of female soap opera viewers in Egypt, where the focus was on the coping strategies of poor or suffering individuals. In the Philippines, religious/cultural beliefs and high-emotion television narratives similarly help provide a sense of 'ontological security' (Giddens 1991; Silverstone 1994, 2005) for sufferers, as we heard Amparo in chapter five say that watching suffering in the news in fact makes her feel 'fortunate' and 'blessed'. However, I took this discussion further by situating people's narratives of everyday suffering within the larger context of economic inequality in the Philippines. Indeed, studies on poverty and suffering tend to succumb to sentimentalism and celebration of coping mechanisms (David 2001a[1976], 42). Listening to the different voices of audiences was crucial in this case to understand the multiplicity of moral codes about suffering in a class-divided society, challenging more essentialist anthropologies of a Filipino morality (Jocano 1997) and pronouncements of a Filipino culture of disaster (Bankoff 2003).

Intersecting Zones of Safety and Danger

The empirical focus on the Philippines has exposed that non-Western zones of danger are themselves stratified from within into hierarchical, and occasionally intersectional, zones of safety and zones of danger. The suffering literature, insofar as it is concerned with inequalities of human life, was brought into productive dialogue with sociological work on class and stratification in this study. Indeed, the concerns about the hierarchies in human life amidst a global backdrop mirror concerns about the 'hidden injuries of class' (Sennett and Cobb 1973) within societies that have festered, and even worsened, as a result of neoliberalism and processes of intense mediation and globalization (Couldry 2010e; Illouz 2003; Ouellette and Hay 2008; Orgad 2009; Wood and Skeggs 2009).

I find that theoretical and psychological elaborations on denial and avoidance in relation to mediated suffering were enriched here by sociological work that links these practices to the broader sphere of cultural consumption (Skeggs 2004; Wood and Skeggs 2009) and ethnographic and historical accounts of interclass relations in the Philippines (Cannell 1999; Johnson 2010; Kerkvliet 1990; Pinches 1999; Tadiar 2004). Linking the suffering and class literatures enriched the analysis of the continuities between upper-class movements of 'flying over' in the metropolis and their surfing of television channels. Chapter three argued that the boundaries between zones of safety and zones of danger are both physical and symbolic and that the relations between the groups who inhabit these are defined by the erection and transgression of class borders in geographic and mediated spaces. This finding expands on current descriptions of denial as a psychological defense mechanism of individuals (Seu 2003) by reconceptualizing denial as a class strategy. Doing so provides impetus for political and ethical critique of structural and institutional forces that discourage cross-class interaction and reproduce social inequalities. Additionally it contextualizes audiences' symbolic practices of 'switching off' within a larger social and ethical backdrop, as I discuss further below.

The relevance of class as an analytical category applies not only to the study of ethics or suffering, but also to the media. Morley (2006) has observed that media studies has drifted away from an early focus on class to a contemporary concern on new 'identities', such as race and ethnicity. However, class in both developed and developing countries is alive and alarming. It should remain a concern of media studies if we regard the access to and exercise of media's symbolic power as a resource for fair and equal (re)distribution (Couldry 2003; 2010), and if we consider the media environment as a 'space of hospitality' for the other (Silverstone 2007). Class has been found as significant in studies of mediated political engagement (Couldry et al. 2007); it should continue to figure as important in reception studies that investigate audiences' interpretations of texts, argued here to be influenced by class positions. It should also remain central in the critique of media representations, insofar as mainstream media 'over-represent' (Wood and Skeggs 2009, 177) or provide 'distorted recognition' (Couldry 2010: 156–8) to those with little resources to represent themselves. As such, this book contributes to these contemporary media studies scholarship on class (Ouellette and Hay 2008; Skeggs 2004; Wood and Skeggs 2009) by foregrounding how people's judgments of value and respectability toward cultural products have an *ethical* dimension: taste judgments are dangerous in how they legitimize or delegitimize particular moral claims being made by sufferers demanding recognition from the wider public.

At the same time, this book advocates for greater intersectionality in future research. The emphasis on class here should not occlude the significance of categories such as gender, age and religion, categories that were also observed to inform moral discourses in relation to television consumption, though to a less intense degrees than class. This book could benefit from a longer discussion of, for example, the process by which commonly shared Catholic doctrines and beliefs about suffering were contracted and splintered along class lines. Class, in the Philippine context, indeed remains exceedingly important, with contemporary work claiming that class greatly shapes various processes – from classed religious practices (Cornelio 2011) to classed forms of racism (Cabañes 2011) to the stratification of sexual identities according to class (Benedicto 2009). This book then argues that analyses of social processes and phenomena in the Philippine context will be severely partial if they fail to examine how these are experienced, contracted and shaped by class.

Instrumental Cosmopolitanism

Whilst the media ethics literature has raised questions of cosmopolitan compassion for distant others, this audience study attests to the continuing primacy of the national in social and ethical issues of inequality, human dignity and responsibility for the other. In fact, I began this research with an interest in cosmopolitanism as a framework to understand audiences' moral obligations to their mediated others (Ong 2009a), but found during fieldwork the continuing relevance and urgency of 'suffering within' and people's (denial of) obligations to their proximal others. Though I found that upper-class Filipinos did not at all lack knowledge of international issues, with some respondents more attentive to international than local news (see chapters three and five), what was concerning was how cosmopolitanism usually came at the expense of regard for the proximal neighbour. Upper-class assertions of cosmopolitan knowledge came alongside ignorance about local issues and stereotypical beliefs about or active discrimination against the Filipino masa. Within the context of the research interview in fact, feigning ignorance about local celebrities is assumed to accrue for the participant higher value by dissociating herself or himself from the *jologs* who lack the correct cultural capital (see chapters three and four).

Certainly, this instrumental form of cosmopolitanism (Ong 2009a: 456–8) would not pass the moral standards of Tomlinson (1999, 195), who demanded cosmopolitans to be able to address social issues in both the global and the local spheres at the same time. Instrumental cosmopolitans, who only desire attachments to global imaginaries whilst 'cutting the network' (Strathern 1996; McKay 2009) or breaking ties with locality or nation, would fail tests of both cosmopolitan ethics (Silverstone 2007; Tomlinson 1999) and communitarian ethics that prescribes help for the neighbour (Clement 1996; Noddings in Smith 1998). This study does not at all intend to argue for the deficiency of cosmopolitanism as a moral project, but what I hope to have demonstrated is how cosmopolitanism can have a double face: as much as it can work as a model for right living, in its instrumental incarnation it can be a righteous assertion of value in one's way of life in opposition to the lack of value in the other's. We have seen in previous chapters that people strategically use the cultural products that media circulate to claim value for themselves, whilst projecting low value to others.

In the case study of distant suffering in chapter five, we also failed to observe cosmopolitan charity for the victims of the May 2008 Sichuan earthquake. I suggested that the absence of cosmopolitan action in that case was influenced more by the (low) quality of visibility accorded to the event by local television than an essential Filipino disregard for distant others, as argued by commentators such as Espina (2010). My study did uncover some cases of upper-class cosmopolitan charity, particularly with those who had more resources, opportunities and interest to do so. Rather than being inherent in Filipino culture, I demonstrated that contemporary and classed moral contests in society shape Filipinos' charitable practices. Chapter five showed how upper-class respondents' cosmopolitan charity tends to be surveilled by the middle-class, especially when perceived as a negation of kinship obligations and when judged as a performance of class superiority. Unlike instrumental cosmopolitanism, the ethical version of cosmopolitanism envisioned by scholars such as Chouliaraki (2006) and Silverstone (2007) for the most part remains an elusive normative ideal when subjected to ethnographic inquiry. This does not mean, however, that this ethical standard should be rejected or replaced, for it offers models of moral media and audiences that are nevertheless desirable – at times even observable.

'Switching Off' as an Undesirable Moral Orientation

Civic and political disengagement in the form of voter apathy or 'switching off' from the news has been criticized as detrimental to a nation's democratic health (Jensen 1995; Putnam 2000). But 'switching off' has also been observed to be the result of social exclusion from the public sphere, and of systemic misrepresentation (Madianou 2005b). In chapter three, I identified 'switching off' as a consumption practice that required ethical evaluation; in my observation of Filipino upper-class 'switching off' from local television, I have argued that it is ultimately an undesirable practice of mediated engagement.

In chapter five, I showed how a general orientation of disengagement disables a person from encountering the diverse moral claims made available in the news – individual representations of suffering that could potentially activate compassion, indignation or charitable action. The fact that upper-class audiences only tune in to the news during times of great crises means that they do not at all witness, and therefore cannot receive the moral claims within, everyday stories of poverty and charity appeals in the news. I also discovered that media charities have been sustained by retail fundraising that consists of low-denomination donations, presumably by low-to middle-income donors, rather than substantial donations from well-off viewers. Whilst it is impossible in the context of this study to trace a direct connection between 'switching off' and media's low receipt of substantial donations, I suspect that there is one. Though audience responses to mediated suffering are complex and unpredictable (expressions of hope, moral justifications, compassionate charity, etc.), there nevertheless exists a possibility for 'effective speech' (Boltanski 1999, 18–19) or charitable action when audiences encounter mediated suffering. Being switched off as a default orientation to the mediated public realm, however, obscures off, any hope for speech or action towards televised suffering.

'Switching off' does have consequences not only for effective speech or action toward televised sufferers but also in the perceptions and interactions with the poor in everyday life. Moving and surfing within worryingly narrow geographic and symbolic zones of safety hinder possibilities of mutual recognition and understanding that might arise from cross-class encounters, whether face-to-face or mediated. Scholars have observed, for example, how everyday interactions across different social groups, at least when approached by the parties involved with a stance of openness to each other, might lead to the breaking down of stereotypes and lead to conviviality (Appiah 2006; Gilroy 2004). Kim (2007) has specifically observed that popular consumption of Korean soap opera in Japan has coincided with greater and positive cultural exchange, such as travel, learning languages, and even interracial marriage. She cites that mediated encounters of the 'other' through consumption of soap opera are actually socially beneficial, especially for these two countries with a long history of political conflict.

Certainly 'switching off' fails standards of 'mundane witnessing' (Ellis 2009) and 'passive engagement' (Donnar 2009), reviewed in chapter one. Basic 'awareness of events around us and of the people who make up our society and wider world' (Ellis 2009, 83) is lacking from upper-class audiences routinely observed to be 'shocked' by televised representations of everyday poverty. 'Switching off' becomes more worrying when coupled with totalized and homogenized judgments about the *masa* and the *jologs*, as I elaborate below.

Cohen (2001, 294) has argued, '"Turning a blind eye" does not literally mean not looking – it means condoning, not caring, being indifferent. Physical vision is a metaphor for moral vision.' In the Philippine case, I have found that, with some audiences, it carries both meanings: not looking at but also not caring about proximal others, only glimpsing the other through blurred peripheral vision.

Classed Moralities

Whilst the media ethics literature has been productive in describing how representations of suffering invite reflection on spectators' obligations to the other, the literature on class and consumption has richly described how judgments of value are made by audiences toward ordinary people on television, particularly in the genre of reality television (Ouellette and Hay 2008; Skeggs 2004; Wood and Skeggs 2009). Linking these two insights was helpful in understanding how class shapes moralities, whilst gaining a better understanding of class and stratification as ultimately moral concerns.

Classed moralities became an important theme in this study upon discovering that judgments of value based on authenticity, respectability and deservedness were made toward televised sufferers. I found in chapter three that upper-class participants brand the majority of the participants within the mediated centre as *masa* and *jologs* – catch-all words of hatred that identify out of fear, anxiety and threat (Skeggs 2005). In chapter four, lower-class participants used visual- and auditory-based criteria to judge those they considered the authentic and deserving poor in *Wowowee*, based on popular stereotypes about skin colour, body shape and accent. Meanwhile middle-class audience respondents asserted moralities of authenticity and resourcefulness to differentiate themselves from their upper-class rivals, whilst co-opting good lower-class values in chapters three and five.

This means that audience responses to representations of suffering are not only shaped by the aesthetic or narrative qualities of the text but are also greatly shaped by audiences' social contexts, the most significant of which was class in the Filipino case. Class shapes audience responses through the constellation of experiences, moral discourses and strategic interests that inform an individual's 'resources for judgment' (Silverstone 2007, 44). In the context of media reception, spectators (dis)engages with media images of suffering by drawing on their personal experiences, habituated judgments of correct value, and media skills in deconstructing texts and their messages.

In this context, morality is less a Durkheimian 'floating mantle over society, pervasive in all of its aspects' (Heintz 2009, 2), than it is splintered, strategically negotiated and reproduced along class lines – a model of morality more in line with the work of Skeggs (2004; Wood and Skeggs 2009), Sayer (2005) and Johnson (2010). This conception of morality is more conflictual rather than functionalist, and more plural rather than singular. This book then argues for the significance of studying morality as a resource differently possessed by individuals (Laidlaw 2002) and whose ownership is linked to access to political, cultural and – perhaps significantly – economic capital. A plausible assumption then is that a less economically unequal society is one that is also less polarized in its judgments of value of human life, or at least one where moral value is less associated with the ownership of the right sort of capitals.

At the same time, by linking media ethics and the sociology of class, we found that the reproduction of inequalities carries real moral consequences. We observed that words of hatred, such as jologs and masa, when evaluated from the prism of the media ethics literature, productively work to deflect any moral claim that may arise in the embodied or mediated appearance of a poor person in public life. These words serve to actively attach negative value to the poor whilst providing a moral justification for disconnecting from them and rejecting any obligation to help.

However, we also found that lower- and middle-class participants actively negotiated their perceived and felt positions of lower value by asserting their own moralities of respectability and authenticity to contest upper-class judgments. Though in some contexts they did wear hegemonic lenses and self-pitifully admitted to having an undesirable status, in many cases they found and claimed value for themselves. My lower-class respondents read off resourcefulness and dignity in the 'media pilgrimages' (Couldry 2003) of sufferers and the emotional labour that they endure in *Wowowee* (see chapter four). These mediated performances of strategic suffering are understood as agentic rather than victimizing, contrary to upper-class perceptions. Successful strategic suffering, after all, entails *pakikipagsapalaran* (jumping at the chance) and careful weighing of the odds, just as it requires knowledge of genre conventions and the rules of the game within the Filipino mediated centre. Meanwhile, middle-class respondents challenged elite dominance by asserting that by being *jologs* media consumers themselves, they were in touch with reality and can relate to what's going on, as we heard Amanda say in chapter three. This illustrates that popular media, rather than simply being trash or completely victimizing their participants (Dancel 2010; De Jesus 2011), are greatly imbricated in everyday struggles for recognition and deserve more careful and more thorough critique and appreciation.

Lay Media Moralities

Just as we observed classed moralities about suffering, we also found classed lay media moralities – assessments of right and wrong media practice – in relation to televised suffering. Dalton et al. (2008) and Seu (2003) found that audiences criticized textual representations of suffering for numerous reasons – the graphic imagery, the evocation of guilt and the lack of authenticity – but argued that these were denial strategies to deflect the moral claims of the text. Similar observations were made here: upper-class respondents recoiled from graphic images, high-emotion appeals, loud sounds and noise, it could be argued, to strategically reduce factual suffering on television to a 'textual game' (Chouliaraki 2010, 120), where the programme 'manipulates you to feel pity', as we heard Happy complain in relation to the controversial *Wowowee* in chapter four.

In addition, the concept of lay media moralities captures in this context a wider judgment of right and wrong that is not only tied to discussions of textual ethics, but also to the ecological ethics of media. Lay media moralities were found to be underpinned by judgments of value about the nature of media visibility and its proper distribution. Lay media moralities were tied to specific ideas of respectability and privacy, and their violation – shame and dishonour – just as they were informed by a respect for the limited resource of media visibility.

Upper-class participants criticized the general convention of over-representing suffering within the mediated centre. Though 74.7 per cent of Filipinos are classified as 'low income', and only 0.1 per cent belong to the upper class (Virola et al. 2013), television practices of catering to less privileged viewers were deemed by my upper-class respondents as an unjust distribution of media resources. Whilst there was sharp critique about the jologs space of television, upper-class respondents nevertheless revealed on particular occasions that they do in fact value some forms of media power: they adore visibility in the respectable genre of news, especially when they are conferred the roles of experts and authorities (and as such they have more control over their representation), and they also fear visibility in contexts where they might be exposed as victims or subjects of scandal or suffering, violating standards of elite respectability and privacy. Where, in other words, they are represented in affectively felt asymmetrical relations of power.

Lower-class informants meanwhile actively pursue visibility, as it is perceived as intertwined with material rewards, given Philippine media practices of 'acting as the government' (De Quiros 2009a). Media pilgrimages and reverse pilgrimages can be potentially advantageous for the lower-class, as these practices typically involve redistribution of various social services. From the media, they have come to expect good judgment, whilst from media pilgrims they demand good faith in the transactions and exchanges carried out within the mediated centre.

Across classes, the shared sentiment is that media power must be correctly distributed and managed. This book attests to how the use and distribution of media power is discussed by ordinary people in moral terms. Their evaluations are nevertheless different from theorists' normative positions of democratizing the distribution of symbolic power (Couldry 2003, 2010e) because audiences are less concerned about social equality than individual, instrumental gains. Privileged viewers criticize media's over-representation of the poor because this translates to fewer opportunities for them to benefit from positive exposure on television; low-income viewers pursue media visibility – in spite of all the physical and emotional labour that pilgrimages entail – for the promise of material rewards.

I argue that future research could benefit from exploring more deeply the diversity of lay media moralities – not only in relation to televised suffering but even in the use of new media platforms – as Gershon (2010) has done in the context of managing relationships through communications technologies. It is important to explore how lay media moralities might be linked to sociological factors of class, age, gender, ethnicity, etc. One limitation of my study has been the minimal focus given to theorizing how lay media moralities arise, develop and perhaps even change because of class or other sociological factors. As Gershon (2010) similarly pointed out, what she calls 'media ideologies' in relation to new media technologies are shaped by knowledge and use of the different new media. Given that my Filipino upper-class respondents used foreign mass media productions (with higher production budgets and with lower quantities and different qualities of representations of suffering) as a benchmark in critiquing local mass media content or practices, they often ended up being shocked, disappointed or even disgusted. How might lower-class audiences' lay media moralities change (or not) if they were afforded greater access to international media? Would they begin to criticize local game shows and reinterpret strategic suffering as actually victimizing rather than possessing agency? Would they question the over-representation of sufferers and the provision of social services? Or would they simply not watch these more distant programmes? These are questions that are worthwhile to ask, but are beyond the scope of this book.

It is important to note that lay media moralities are constitutive of media criticism at the most basic, local level. Though we found that these were less informed by supererogatory first principles and instead driven more by strategic interests and shaped by their habitus (Bourdieu 1986), I believe that recording lay media moralities enabled a deeper understanding of the social consequences of television in Filipino society. I also argue that attentiveness to lay media moralities provided an opportunity for sufferers or others to be heard in relation to how they actually experience *media* and regard their own televised representations. Moving forward, I believe the media ethics literature can benefit from a research agenda that presents the points of view not only of producers and donors (e.g., Seu and Orgad 2010) but also of those represented by producers.

In the next section, I ask what the discovery of multiple and contesting lay media moralities in the context of televised suffering means for media ethics. Are there better or worse lay media moralities between those who are normally switched off from local television and those of the *masa*, who are actually more knowledgeable about local media conventions, but might be 'too close' (Silverstone 2007, 48) to their media patrons?

An Anthropological Ethics of Media

This book responds to the call for an anthropological ethics of media (Born 2008). I have summarized how this approach was able to describe how abstract philosophical norms (i.e., responsibility and cosmopolitanism) are translated into actual contexts of media production and reception, and how an audience media practice of 'switching off' within a context of social inequality can be ethically evaluated. Now I turn to how this approach has been useful to inform ethical critique of media production.

Given the plurality of lay moralities and lay media moralities in class-divided Philippines, an anthropology of moralities can, in theory, provide a relativist position and assert that a media programme or genre escapes moral judgment. As Heintz (2009, 5–7) has reviewed, the anthropology of moralities is historically divided on whether anthropologists should prescribe moral judgment or only describe their subjects.

However, through an ethnographic approach, we were also able to evaluate holistically a particular process of mediation, in this case that of suffering. By tracing different moments in the mediation process and attending to ethical concerns that arise from each moment, an anthropological ethics of media, I argue, could actually provide more careful ethical judgment of media production.

In chapters four and five, we found that locating the ethics of televised suffering within the text, particularly the specific debate of whether to represent suffering as having agency or no agency, leads to a difficult quandary: whose interpretation of agency is correct – the researcher's or his informants'? If the latter, which informants are then correct, given that they have different interpretations of agency and subsequent discourses of compassion and disinterest? Is Red correct that transactions within the mediated centre are often *gamitan* and 'win-win situations'? Or should we believe Natasha who says that television is often 'abusive' and 'has a chokehold on the *masa*'?

Chapter four particularly argued that ethical analysis of the programme *Wowowee* must consider the multiple lay moralities of audiences and acknowledge the real and substantive gap that this show fills in Filipino society as a legitimate mediated space of recognition and redistribution. Rote arguments of media exploitation, harassment and dumbing down of audiences, whilst possibly driven by the noble intentions of engaged media criticism, further totalize and victimize the audiences whose poverty forms the basis for the show's moral community. Ethnographic inquiry challenges conventional media criticism that television has totalizing, infantilizing and victimizing effects on audiences, because it can expose its consequences from presenting the perspective of audiences who have the most critical and emotional engagement with the show. Interviewing actual audiences revealed that lower-class audiences possessed greater knowledge of local genre conventions and media practices that contextualized their affective consumption of the programme, whilst upper-class audiences mostly drew from their knowledge of foreign genres and their habituated taste judgments to criticize the programme. Whose lay media moralities are correct in this case: those with historical knowledge of the programme or those with referential knowledge of foreign production values and conventions?

An ethnographic perspective expands conventional focus on textual ethics of mediated suffering (and its focus on the agency/victim debate) and demands consideration of ecological ethics (how media interact with sufferers, how media over-represent or democratize access to the mediated centre, how media are responsible or not in causing deaths of media pilgrims, etc.), as well as audience ethics (how and why audiences 'switch off' or tune in, express compassion or donate).

With this holistic perspective of the mediation of suffering, I argued that an anthropological ethics of media could provide moral judgment of *Wowowee*, the news, and general practices in the mediated centre by identifying and considering these questions that emerged as relevant:

- Do they contribute to lessening or amplifying social divides in society?
- Do they provide cross-class recognition rather than over-representation or segregation?
- Are they hospitable or not to multiple and diverse claims of suffering?
- Do they engage or not television audiences in different social positions and having different resources for judgment?

These additional questions were important to ask in the Filipino case. They may be equally significant questions to consider for ethical media criticism in other cultural contexts.

Whilst media ethics scholarship about suffering rightfully critiques media content that represents suffering, a critique of production processes and interactions between the media world and ordinary people is equally necessary. I find that the philosophical and text-centred media ethics literature can potentially benefit from a dialogue with studies that pay ethnographic attention to media policies and procedures of interacting with ordinary people (e.g., Ouellette and Hay 2008; White 2006; Wood and Skeggs 2009), treating employees, (e.g., Hesmondhalgh and Baker 2010; Gregg 2011) and engaging with public criticism. After all, concerns about agency and victimhood are concerns about exploitation and empowerment; these issues certainly arise not only from textual representation but also from processes of interaction between media actors who wield power and ordinary people who often do not. Linking the media ethics literature with more materialist approaches that examine people's direct experiences with the media frame (see also Couldry 2000, 2003; Madianou 2011) is a productive step towards a more holistic critique of the misuse and abuse of media power.

A holistic ethics of mediated suffering that situates criticism of particular texts alongside the wider practices and institutional frameworks that make up media cultures could better identify the real consequences of media practices on their audiences, provide more targeted or culturally relevant recommendations for media regulation, ask more particular questions in ethical critique of media and continue to spur necessary debates about a universal or global ethics of media and particular or anthropological ethics of media (Born 2008; Couldry 2008a; Silverstone 2007; Zelizer 2008).

Mediating Suffering, Dividing Class

This book contributes to the corpus of scholarship on mediation (Chouliaraki 2006; Couldry 2000, 2008b; Livingstone 2009; Madianou 2005b; Silverstone 2005), which acknowledges the dialectical relationship that media have with society – shaping and being shaped by the context in which media are produced and consumed. This thesis affirms this dialectical process of how suffering as a social experience is transformed by television, just as television is informed by the social and cultural experience of suffering. I have used the categories textual ethics, audience ethics and ecological ethics to analytically fix media ethics concerns to moments in the mediation process, whilst arguing for the need to see their interconnections.

This book asked, 'How do audiences in their different contexts respond to televised suffering?' The concise answer is that audiences' responses are informed by their class positions, just as responses are influenced by the media texts, genres and practices that make class by reproducing and amplifying class differences.

We observed in this book that the mediation of suffering is a process imbricated in moral contests. First of all, suffering in itself is a moral issue; responses of other-directed compassion or inward self-reflection were observed in mediated contexts of moral dilemmas similar to but also different from face-to-face encounters with suffering. Second, morality is imbricated too in judgments of compassionate, just and 'correct' distribution of media visibility; media power is considered as a valued resource and culturally expected to be used in recognizing and over-representing the Filipino poor. Third, the mediation of suffering provokes multiple classed moralities – moralities in contests of legitimizing and justifying particular audience responses to suffering and particular media practices. Denying obligation, giving to charities, expressing compassion, 'switching off', 'coping mechanisms' (David 2001a[1976]) and criticizing or appreciating television programmes are the multiple audience responses contested along classed divides.

Following the media ethics literature, this book attests to mediation as a practice of moral concern. Following Chouliaraki (2006), the mediation of suffering is about television's selective ability to provide moral claims to respond to certain disasters over others; following Silverstone (2007), it is about television being hospitable or not for minorities consistently denied ability to represent themselves.

In a context of everyday poverty, however, the mediation of suffering is all these but more. A representation of disaster does not only present, or fail to present, a claim to act or not on this or that disaster. Every representation here serves a dual purpose: it serves as a singular moment by which suffering can be individually recognized and acted upon; yet it also exists *in* continuation and as an extension of everyday social relations between rich and poor. News about fire or flood not only mediates audiences with events of fire or flood, but also mediates audiences with neighbours in perpetual proximity to fire and flood. A news clip or a factual entertainment programme invites or does not invite compassion for an event of suffering, but it exists in continuity with historical relations between rich and poor. In other words, representations of suffering contribute in the making of class by amplifying or minimizing class divides. In this light, the mediation of suffering is then a moral concern that implicates the individual spectator within a wider social context of conflict and 'flying over' (Tadiar 2004), and at other times *damay* (mourning, sympathizing) and cooperation between those who have and those who have nothing (Cannell 1999).

We found that the mediation of suffering is a complex and holistic social process. News and entertainment programmes offer different forms of recognition: news presents diverse moral claims from sufferers near, far, individual and collective; entertainment gives voice to selected sufferers within playful, though intensely surveilling, games of chance. Though different in many ways, both genres are also remarkably similar in their address of the Filipino poor and the material rewards that they promise to bestow upon them. Across television genres, suffering is mediated symbolically and – for the 'lucky' or 'deserving' few – materially. In the Philippines, the donors and volunteers to mediated appeals are not only privileged audiences at home – as is the typical assumption in Western literature: here, even low-income viewers and the media themselves serve as donors and volunteers.

Though the Philippine mediated centre is not at all a popular space among the upper class, what is curious is that it is nonetheless owned by old rich families (McCoy 2009), and its operations are managed by upper- and middle-class employees. Upper-class viewers' lack of participation in this centre aside, we can talk about the mediated centre at its most basic as always-already a space for cross-class interaction. Programmes are produced and written by upper- and middle-class owners and producers[hosted and starred in by upper-, middle-, even formerly lower-class celebrities; and always geared towards the *masa*, whose needs and sensibilities are anticipated. The success of media as business ultimately depends on the quantity of viewers who remain loyal, even dependent, on television networks. As much as media producers aim to mediate suffering then, its mediation is perhaps constrained and could not in practice be completely transformative.

So, though it is occasionally commendable that the media 'act as the government' (De Quiros 2009a) in a country of everyday suffering and weak public institutions, the media's will to address suffering cannot but be constrained by its contradictory interest to profit as a private business. Though we have seen in previous chapters that media power in the Filipino case is more overt than in other countries because of its exercise of material redistribution beyond media's typical symbolic recognition, the mediation of suffering is not completely transformative. Despite the forms of agency that audiences display in their different interactions with media, ultimately class is made, reproduced and amplified in narrative content of media representations and processes of pilgrim/producer interaction in the mediated centre.

Because mass media require mass audiences, the mediation of audiences' suffering cannot but be gestural rather than structural. It cannot but follow the logic of dole-out charity rather than long-term-oriented assistance or development. This charity refracts cultural and historical structures of personalistic patron-client ties (Kerkvliet 1990), and so likewise demands audiences' obedience and loyalty in their participation and viewership, as we saw in media's peculiar form of moneylending (see chapter four). Following this, media recognition and redistribution cannot but depend on a politics of pity (Arendt 1963; Boltanski 1999), with its emotional and entertaining registers of compassion, rather than the impartial distribution of rights to voice and visibility. Political economic interests of media constrain them to act so, not only in the Philippines, but almost universally (Couldry 2006; Silverstone 2007).

However, though media and their audiences are in highly asymmetrical power relations and though media and audiences hold varying and competing interests, the previous chapters nevertheless attest that media's provision of emotional and material comforts and general practices of over-representation (Wood and Skeggs 2009, 177) remain of genuine significance to the poor. Over-representation, though derided by local elites as a shameful 'washing of dirty linen' (Cordova 2011), provides comfort and reassurance, even a sense of belonging to a moral community of sufferers, against their everyday experiences of exclusion and rejection.

The study in which this book is based found many things peculiar to Philippine media and their audiences. But it also discovered and reflected on issues of significance to other empirical contexts. Philippine television, though unique in its formalized synergy of charity with factual television genres, shares basic similarities with Mexican (Garcia Canclini 2001; Pertierra 2010) and Greek (Madianou 2005a) media systems in its occasional provision of social services in the context of weak welfare state systems. Future comparative work can more closely examine similarities and differences in media ethics concerns of exploiting or victimizing participants across these contexts. Future work can also link these ethical debates with wider practices of media regulation in different countries. Though there is no intention of imposing legal frameworks from one culture upon another, just as I refrained from doing so with regard to normative philosophical frameworks, comparing insights from different cultural contexts can inform public debate about media responsibility and regulation. In the Philippine case, there has been lax media regulation and accountability, even during the exceptional tragedy of a game show stampede (see chapter four) and interference during a hostage crisis (Ong 2010; see chapter five). Locally, Couldry's (2006, 101) observation similarly applies: '"media independence" [is used] as a shibboleth to render illegitimate any outside moral or ethical scrutiny' alongside routine pronouncements of 'Is it a sin to give people hope?' (Isis International 2006). We therefore require continued media criticism that demands greater accountability from an institution whose intervention in social suffering has symbolic *and* material consequences.

It was my original aim to dialogue with the media ethics literature. By recording voices of sufferers themselves, it was my intention to show that there is much to learn by talking to those whom we wish to help and empower through our scholarly interventions. Listening to people who identify as sufferers themselves led to the discovery that they are not as preoccupied as we scholars have been as to whether television represents them as active agents or needy victims. How they read off agency from their own media pilgrimages differs from how privileged others in society do and likewise differs from theorists whose normative standards might be violated in such interactions. Rather than slip into easy relativism, however, I believe that the challenge for media ethics scholars is to engage with and exceed the agency/victim debates by considering the holistic social and ethical consequences of mediation, by considering some of the questions I have suggested above, as well as by considering the particular manifestations and practices of media power as situated within local contexts.

I also hope that this book has emphasized the value of audience studies and its traditional strengths in exposing the plural and contesting social groups and identities in contemporary society. Audience studies enabled us to consider the different uses and consequences of media to people. By interviewing different groups of audiences, we saw the profound significance of ethical debate about mediated suffering: television provides opportunities to address particular catastrophes, but also helps or hinders social relations within cities and nations. Had I interviewed only affluent viewers (the donors) or only sufferers (the recipients), I would likely have come up with a different ethical critique of media programmes and practices, just as I would likely have arrived at a different conclusion about audiences' (un)desirable television consumption practices. I would have also failed to understand how personal value and morality are dynamically and relationally contested along class lines that television often exacerbates rather than mediates.

Postscript

As with most media research, this book captures a particular moment of a rapidly changing and adaptive media environment – and their ever-fickle, variably engaged and occasionally elusive audiences and users. Between the last month of fieldwork in January 2011 and the completion of this book, the Filipino media landscape is in the midst of a gradual but nevertheless dramatic reshaping not seen since the post-martial law period of media liberalization.

Today I would say that Filipino television – and the broader mediated public sphere that it is part of – are redefining the terms of mediated recognition of suffering and the participation of ordinary poor people in symbolic spaces. Whereas the story told by this book has been the over-representation of poverty, and the classed contests of value that play out from these excessive spectacles, now Filipino media are at a turning point of refining and curtailing the excesses of the poverty of television in favour of the more aspirational imagery of middle-class Filipino lifestyle and culture. Mass media content, long-criticized for melodramatic narratives of poverty akin to 'feeding baby food to adults without teeth' (Dancel, 2010), now attends to the excesses of luxury in fictional (e.g., TV drama *Magkaribal*) and factual media (e.g., the transition of Manila socialites from the scene of nightlife cultures premised on concealment to the screen of tabloid/reality TV cultures premised on perpetual display).

In social media, interactional grammars increasingly demand fluency with proper middle-class cultures and moralities (hence you become surveilled as *jejemon*), where people celebrate access to exclusive and aspirational spaces; even journalists and politicians dependent on *masa* followers remake their celebrity by representing their social distance from the *masa* through their Instragramming of luxury holidays. Critics today now worry how political protests orchestrated by Filipinos in social media seem to involve very little participation from the working class yet gain disproportional visibility in traditional and social media thanks in part to Instagrammable photo-ops that are oriented towards sharing, commenting and – eventually – easy remediation by mainstream news (Curato 2013).

Even in the context of natural disaster, the quality of representing suffering seems to be in the midst of change. In the aftermath of Typhoon Haiyan, which claimed over six thousand lives in central Philippines in November 2013,

I would argue that Filipino media – and social media most particularly – challenged the representations of atrocity 'at its worst' (Orgad, 2008) that circulated in global news. Shifting the focus away from (perceived) Western media's fixation with 'dehumanized' suffering bodies, unruly survivors desperate for aid, and an inept third world government failing its people, middle-class dominated social media turned their lenses toward themselves through their selfies as 'voluntourists' in disaster zones, fetishized Western male bodies in their photo-essays about muscular and desirable superhero volunteers from foreign aid agencies and made viral inspiring and heartwarming narratives of resilient survivors who help themselves. Even social critique of government, privatized recovery and media coverage during and after the disaster was surveilled ('Stop complaining! Just help!' in Get Real Philippines 2013; Paul's Alarm 2013; Ramos-Aquino 2013) in line with middle-class moralities of saving face, maintaining dignity and depoliticizing charity. These middle-class and elite discourses were adopted as official discourse by government. The glossy state-sponsored tourism campaign called #PhThankYou was launched just three months after Haiyan with the objectives of thanking generous foreign publics who donated to Haiyan relief efforts and then inviting them over to the Philippines as tourists visiting an exciting, fast-developing country. Consistent with middle-class moralities of positive imagery of suffering, the campaign involved a video that juxtaposed rehabilitation in disaster zones with middle-class spaces of leisure, fusing the most refined vocabularies of tourism and humanitarianism in a way that jarringly interrupts the pessimistic narratives of post-disaster recovery that circulate in Filipino and global news (see Ong 2015b).

To me, these shifts in representation can be partially attributed to broader shifts in Filipino economic, social and technological landscapes. First, the Philippines' expanding economy, said to be second only to China in terms of growth (Gatdula 2014) with middle class driving the market consumption according to the latest global survey of AGB Nielsen (Dinglasan 2013), reconfigures the 'sachet economics' business model and moves away concern for DE markets towards the middle classes: media content then follow the lead of their advertisers and the markets they attempt to capture and remake. Second, middle-class Filipinos' 'worlding' aspirations for positive global recognition are stoked by optimistic news of the Philippines as an economic miracle in the context of global financial crises after decades of negative labeling as the 'sick man of Asia' and 'nation of servants'; this provides impetus for the reimagining of the country and its established culture of disaster in glossy imagery and hopeful vocabularies of suffering, coopting and refashioning original meanings of resilience for instrumental gains of the middle-class. And third, the Philippines' seeming embrace of digital technologies and particularly social media have contributed to the dramatic reshaping of the Filipino mediated public sphere. The Philippines' topping of global rankings as 'number one social media capital in the world' and its city of Makati as 'number one selfie capital in the world' according to *Time* magazine (2014) have dangerously led official and mainstream media discourse to increasingly follow social media discourse in spite of continued digital divides. Statistics agencies indicate that internet penetration remains at 36 per cent, from 32 per cent in 2000 (Albert 2013), suggesting that social media voices are predominantly urban middle-class voices, increasingly relied on by Manila-centric traditional media to speak for the poor, rural, suffering voices in disaster zones.

These changes signal to me a gradual move in Filipino media to emulate Indian media, criticized today as being in denial of poverty in their disproportionate attention towards the cosmopolitan dreams and aspirations of its middle-class (Mankekar 1999; Sainath 2009). This shift would mean a new crisis of representation in the mediation of suffering, where the emerging poverty of television lies not in its traditional excesses, shock effect and noisy sensationalism but in its soothing, eerie silence and absence.

APPENDIX

This book draws from a four-year PhD project carried in the Department of Sociology, University of Cambridge and fully funded by the Gates Cambridge Trust from September 2007 to July 2011. Ethnographic research with television audiences, supplemented with interviews and participant observation in three television networks, was conducted intermittently from June 2008 to October 2009, June to July 2010 and December 2010 to January 2011.

This appendix summarizes the project's methodology and presents information about the background and sampling of project participants and the general conduct of fieldwork. The ethnographic approach taken here draws from the tradition of audience ethnographies in media studies and the anthropology of moralities in the attention to lay moralities expressed in ordinary people's media talk about poverty and suffering on television.

Ethnography of Media and Moralities

Ethnography is traditionally defined as the use of participant observation requiring long-term immersion in a particular culture. Hammersley and Atkinson (1995, 1) define it as participating 'overtly or covertly in people's daily lives for an extended period of time, watching what happens, listening to what is said, asking questions – in fact, collecting whatever data are available to throw light in the issues that are the focus of research.'

In media studies the use of ethnography is well documented. Within the 'ethnographic turn' of media research in the 1980s to 1990s, studies highlighted how media artifacts are appropriated in people's everyday lives and how media messages are differently interpreted by audiences (e.g., Gillespie 1995; Radway 1985). An ethnographic perspective is understood as well suited to examine the 'double articulation' of media as texts and as technologies or objects (Silverstone 1994). This holistic approach not only examines processes of *reception* in investigating specific interpretations of media messages; it also accounts for general processes of *consumption* in studying what people do with the media in their everyday lives (Morley 2006).

Ethnography is understood to be ideal in studies that employ the framework of mediation in examining the 'circulation of meaning' in media and social life (Silverstone 1999, 13). Ethnography is sensitive to different moments that constitute the mediation process, namely media production, media texts, media technologies and the social and cultural context (Madianou 2009). As my study is concerned with the interrelationships between media messages, people's responses to these messages, and the social and cultural factors that shape people's responses to these messages, ethnography allowed for a widened remit of analysis than traditional methods of interviewing. This approach is consistent with ethnographies of media which avoid media-centrism by providing thick description of audiences' domestic and social life beyond the media (Abu-Lughod 2002; Gillespie 1995; Kim 2005; Madianou 2005b; Mankekar 1999).

At the same time that the study followed contemporary ethnographies of media, my research also drew from perspectives elaborated in the sub-field of anthropology called the anthropology of moralities. As discussed in chapter one, the anthropology of moralities examines lay normativities or lay moralities in specific socio-historical contexts. The project then approached the domestic context of watching television as in itself evocative and provocative of lay discourses of good or bad and right or wrong (Howell 1997; Heintz 2009). Whether this experience of seeing other people's suffering is regarded as a 'morally problematic scenario' (Heintz 2009) or not was the starting point of my ethnographic inquiry. By probing respondents' expressions of obligation, compassion, guilt, etc. – and supplemented by participant observation – I tried to identify local moral codes within which people act or are expected to act, bearing in mind categories of class, age, gender, as well as the qualities of media texts and genres that depict poverty in Philippine television. It must be noted that the ethnographic exploration of lay moralities has clear resonances with the spirit of Aristotelian *phronesis* developed by Lilie Chouliaraki (2006) within the literature of media ethics. While Chouliaraki examines texts as enactments of ethical discourse in society that themselves provide occasion for moral reflection and action from audiences, this study looks at audience responses and practices as themselves shaped and co-constituted by the texts that they encounter.

Project Respondents

I used purposive sampling (Lindlof 1995, 126) to obtain a good balance of respondents in terms of class, gender, religion and age. As reviewed in chapters one and two, previous studies have highlighted the significance of class, age and gender as shaping people's moral reasoning, prosocial behaviours and

Table 4: Respondent Characteristics (Total: 92)

Female	50	Catholic	68
Male	42	Christian-Other	9
		Muslim	2
Under age 20	15	Non-practising	8
Age 20-35	33	No religion	5
Age 36-60	37		
Over age 60	7	Volunteer/Member of Charity	16
Upper-class (AB)	30		
Middle-class (C)	26		
Lower-class (DE)	36		

discourses about distant others (Dalton et al. 2008; Gilligan 1982; Hoijer 2004; Kyriakidou 2005). And in the Philippines particularly, I have argued that class is a central category for social analysis, as discussed in chapters one and three.

In total I had 92 respondents (see Table 4), 30 of whom became regular respondents whom I visited multiple times.

Though I was concerned with achieving roughly equal numbers of respondents to satisfy categories emphasized in the literature, I also aimed for a maximum diversity of responses in order to capture the diverse experiences of life in the city and diverse experiences with media.

My recruitment of upper-class respondents was primarily enabled by my part-time lectureship at the Ateneo de Manila University, a top-ranked university in the country that is known for catering to elite Manila society. As a Jesuit-run institution, religious formation, outreach and public service (to be 'persons-for-others') are highly emphasized in the curriculum, where participation in build-a-house and literacy programs are compulsory for graduation. At the university, I was able to interview students, parents, older alumni and staff (including middle-class clerical workers and lower-class labourers).

Through my Ateneo contacts, I was able to meet with volunteers and employees of faith-based and charity organizations, such as *Gawad Kalinga* (Bestowing care) and Ayala Foundation, which are popular among upper and middle classes.

Visiting these groups also allowed me to network with people who had contact with so-called depressed – or slum–communities. On several occasions, I joined project managers in visits to squatters' areas, where

I was able to make contacts for later interviews. The slum community that became a primary fieldwork site in meeting people from the lower-class is Park 7. Park 7 is an 'invisible' squatters' community of roughly 100 families and is a few minutes' walk from Ateneo de Manila. I say 'invisible' because the entrance to Park 7 is a nondescript gate that could be mistaken for a side-entrance to the upper/upper-middle-class Xavierville neighbourhood that occupies much of this space. Entering the gate to Park 7 is entering a sprawl of narrow alleys divided by a wide creek known to cause flooding during typhoons. Park 7 is where I managed to interview self-confessed fans of ABS-CBN, self-identifying as *Kapamilya* (Of one family) – the tagline of the network. Some of these heavy viewers of television also shared experiences of being studio audience participants in game shows, which challenged me during fieldwork to think deeper about how people's direct experiences with media might affect their interpretations of televised suffering. My respondents in the Park 7 community were usually older women, so, to fill the gap of younger lower-class respondents, I felt compelled to add another contact point.

During fieldwork, I also volunteered as teacher trainer in a public school in the Mandaluyong area. This university is vocation-based and specializes in nursing, electrical engineering and work in service industries. Here I conducted fortnightly classes on transferrable skills such as the use of PowerPoint and other teaching tools for university staff. In the process of work, I gained access to students and staff. I was able to interview students from ages 19 to 40 who balanced school with day jobs as nannies, drivers, hairdressers or technicians or with night jobs as call center agents and fast food service staff. By doing participant observation and engaging conversation at a canteen near the university, I came to interview staff and students. The canteen's television was turned on at all hours, and I would often come to visit during noontime when the programme *Wowowee* was on.

Following the principle of maximum diversity, I also used my professional and academic contacts to interview people whom I thought might provide a different perspective from existing respondents. I interviewed people who were victims of calamity, such as family members of victims of a sunken passenger ferry. Some of these respondents also had direct contact with the media by being interviewed for news and public affairs programmes. These interviews probed the kinds of assistance that television networks provide in contexts of disaster, informing my analysis of the mediation process.

In September 2009, when Typhoon Ondoy hit Manila and submerged 80 per cent of the city, I also made sure to follow the trail and revisit old respondents affected by the disaster and meet new ones.

Whilst having interviewed over 90 people with various backgrounds, I do not claim representativeness. Representativeness is not the main concern of qualitative research (Miles and Huberman 1994), but what matters most in ethnographic studies is for the multiple contexts of audiences to be approximated in order to 'maximize what we can learn' (Stake 1995).

Methods and Fieldwork Milestones

My ethnographic project relied on several methods to explore empirically the mediation of suffering: the group interview, the life story interview and participant observation.

Group interview

The group interview format followed general principles of reception research in media studies. But whereas reception studies traditionally investigate people's agreement or disagreement with ideological messages, typically in the social setting of a focus group interview (Morley 2006), following the anthropology of moralities, I employed the reception interview to elicit moral discourses from respondents. Similar to the approach of moral dilemma elicitation (Heintz 2009), I probed people's expressions of right and wrong and good or bad in relation to selected television programmes, especially in their depictions of natural disaster, death and mundane poverty.

The group interview was the method that I relied on in early stages of fieldwork that focused on people's responses to the Sichuan Earthquake in China and Cyclone Nargis in Burma. This round of interviews consisted of focus group interviews with three to six participants each. Respondents knew one another as friends or neighbours. The cover story I used to explain the rationale of the interview was that I was conducting research on media habits of Filipinos of different backgrounds. This decision to have a cover story for this round of interviews I took after Norgaard (2006), who probed people's (lack of) interest in environmental dangers in her ethnography in Norway. Norgaard argued that providing a general rather than specific cover story for the interview enabled her to probe people's awareness and civic engagement whilst avoiding perils of socially desirable or pre-empted answers among her informants. This advice was helpful for this study, as I attempted to minimize expectations that they should display awareness about a humanitarian crisis in China in the social context of a group interview.

Using a semi-structured approach, the eighty-minute audio-recorded group interviews began by asking basic demographic information from the respondents. Then general questions about media consumption were asked: I probed favourite television networks, programmes, genres and celebrities as well their use of other media such as radio and the Internet. I also asked about their knowledge of and interest in foreign and local issues in the news. I then asked about the Sichuan earthquake and other tragedies in the news. I probed what sources of information they had and how accurate their statements were, say, in relation to the number of casualties or the reported government response. I also listened carefully to emotional expressions of compassion, anger and guilt previously identified in the literature. And then, I presented three preselected news clips of the earthquake from ABS-CBN's *TV Patrol World*. I prompted discussions about what they thought and felt about the clips. The final set of questions provoked reflection on textual elements of the clips (what they thought about uses of close-up or music or the reporter's manner of asking questions) and the media themselves (what they thought about the general conduct of media during disasters, toward the poor, and their general usefulness or importance to society).

The data that I gathered from this first phase of group interviews helped me draw out particular patterns and themes that would prove resonant throughout my study: class-divided media consumption patterns, emotional expressions towards sufferers, judgments of media practices and different justifications of action or non-action toward sufferers. However, I also found that the format was not conducive to probe certain themes that emerged from the interviews but could not be explored in a group format and/or the predetermined focus on particular news clips of suffering. For instance, a common occurrence during interviews was comparative references to distant versus proximal and personal suffering. Discussions of mediated suffering consisted not only of expressions of different discourses of compassion for televised sufferers as previous studies have identified (Hoijer 2004), but involved retellings of personal experiences of loss and tragedy, observations of media exploitation, and judgments about what sufferers 'ought' to do. These statements were used to explain why they felt compassion or not for particular sufferers. In order to probe all these contextual data then, I employed life story interviews and participant observation with my respondents in their everyday lives.

Life story interview

As mentioned in chapter one, the life story interview is important in the anthropology of moralities as it is argued that, in the process of narrativizing an individual's life choices and experiences, moral beliefs and values are not only uncovered but provoked (Heintz 2009). Life story interviews were used in the study for three reasons. First, life story interviews provided me with narratives of moral dilemmas in people's lives (including the decision-making involved in donating or volunteering for charities), shedding light on how they negotiated abstract moral values in their everyday lives. Second, they allowed contextualization of people's responses from group interviews, providing material for comparison and triangulation. Life story interviews, I found, enabled me to gather data about more social, cultural and even biographical, factors that might affect moral discourses and practices, as respondents talked about how family, school, religion, even travels abroad, contributed to certain values they held about compassion, responsibility, human dignity etc. By decentering the media from analysis, life story interviews allowed me to think about what past experiences and wider social and cultural factors that come into play in the very moment an individual encounters suffering in the news or a reality show. And third, the individual nature of life story interviews allowed respondents to express certain moral values and judgments that they were unable to in group settings. Personal confessions about, for instance, disinterest with foreign news or disdain for reality show participants were sometimes expressed in the life story interview rather than a group interview where participants feel judged by peers.

For the life story interview, I drafted thirty respondents. I selected twenty-four of thirty respondents from the group interviews, where they had offered unique and articulate perspectives or narrated personal experiences that I found significant to probe in private discussion. Six interview respondents I drafted with the specific intent to follow new leads and create maximum diversity (such as victims of calamities, volunteers in charity groups or student leaders).

These interviews were less structured than the group interviews, as the personal subject matter also demanded a more personalized approach in encouraging trust and openness. The life story interview typically began with an inquiry into the respondent's childhood and family life. I asked about the moral values respondents held important and the sources of these values (family, school, religion, media etc.). Following the idea of moral exemplars (Heintz 2009), I also asked about their role models – whether in the private or public spheres – and asked what values they admired in these people. I also probed people's understandings of suffering and offered them to give definitions and examples. I elicited evaluations of specific virtues previously identified in the literature on mediated suffering, such as compassion and responsibility. I also probed what they thought about local norms such as *hiya* (shame), *damay* (mourning) and pity, as well as Catholic teachings on human dignity, humility and the redemptive value of suffering. I also probed what people thought about the link between obligation and distance – whether they felt that people are responsible to help suffering others they were not related to and are so far away. I also probed perceptions of class divisions in Philippine society. Among upper-class respondents, I asked about their interactions with the poor and their perceptions of them, comparing theoretical discussions of agency and victim with lay moralities of respectability and resourcefulness. Among lower-class respondents, I similarly asked about their interactions with and perceptions of the rich. My final set of questions dealt with the media: whether they had direct interactions with the media and what they perceived are media's positive and negative contributions to society. The data in chapter three, which focuses on people's media consumption habits and aversion/affinity to stories of suffering, were primarily derived from life story interviews as supplemented by participant observation and group interviews.

Participant observation

Participant observation was done in various settings. In the context of life story interviews, especially those that were conducted in respondents' homes or schools, participant observation was carried out in the form of taking note of people's material possessions, styles of interaction with friends/strangers and presence/absence of different media technologies. Many occasions of participant observation involved visiting respondents in private (such as their homes) or public settings (such as the school canteen), watching television, taking note of conversations and posing questions to the people present.

There were also occasions when the remit of analysis did not involve the media at all. Among upper- and middle-class respondents, these contexts were charity group-organized visits to depressed communities, fundraising dinners and concerts, school rallies and church visits, among others. Among lower-class respondents, participant observation was carried out by hanging out at the town hall and basketball court of Park 7, church visits and hanging out in the mall, among others. In latter stage of the study, I also visited evacuation centres in hard-hit areas of Typhoon Ondoy (Marikina and Bagong Silangan) as well as relief centre sites in Ateneo, ABS-CBN and GMA. Data gathered from participant observation allowed me to gain a more rounded picture of particular respondents, as it was through this method that I was able to meet the people that they found important and responsible for. In these natural settings, I was able to observe how they relate with friends and also how they talk about and relate with various others.

Additionally, I conducted participant observation in contexts not involving media audiences. Through professional contacts, I made repeated visits to television networks, specifically their charity offices and programming departments where I conducted expert interviews and observation of behind-the-scenes production processes. I attended guided tours of television studios and participated as a studio audience member in selected talk and reality shows. I also managed to interview producers and public relations executives connected to news programmes and *Wowowee*.

The holistic approach of ethnography allowed me to confirm assumptions, follow new leads, meet new people and have a deeper, more intimate knowledge of and relationship with my respondents. For instance, by comparing responses in group interviews and life story interviews, I observed how suffering was generally interpreted by (some of) my respondents as poverty. Specifically, in group interviews about news and entertainment programmes, respondents reserved emotional expressions of pity for people they considered poor. And in life story interviews, when prompted by direct questions about definitions or exemplars of suffering, people responded by talking about 'beggars on the street' or, very generally, 'the poor', 'the homeless' and the '*masa*' (masses). I was also able to get a deeper understanding of certain respondents through interactions in different contexts. For instance, one student whom I thought was unsympathetic – at least in a group interview about suffering on entertainment television–expressed sentiment for sufferers in a news programme on another occasion. This discrepancy hinted that her responses were underpinned by latent evaluations of different genres rather than a generalized compassion fatigue, contempt for the poor or mistrust of all media – assumptions that I could have made had I used only the data from one group interview.

A note on textual analysis

As previous researches on media and suffering have predominantly employed textual analyses, I made a conscious decision to take an audience-centred approach in this study. That is why, in the book, descriptions and analyses of text are carried out side-by-side with and contextualized by audiences' responses to these texts. This approach is consistent with the anthropology of moralities, which refrains from uncritically imposing Western or universalist theoretical prescriptions on local phenomena, not to mention the researcher's own moralities. In this context, the textual analyses of *Wowowee* in chapter four and selected news clips of local and distant suffering in chapter five are informed by categories derived from the textual ethics literature. The impressionistic analyses of general media and generic practices are also informed by insights identified in ecological ethics, particularly the debates about the ethical recognition versus the exploitation of invisible and voiceless others in mass media. At the same time, familiarity with Philippine television history, regulation and political economy as well as generic conventions was useful in analysing texts and informing judgments of victimhood or exploitation. The expert interviews that I conducted also helped here to provide insight to industry practices and conventions.

Knowing specific programme content as well as generic conventions were necessary in analysing why my participants responded the way they did toward particular news clips, entire programmes or even entire genres. At the beginning of fieldwork, it was necessary that I became familiar with the television programmes that my respondents followed – even though some programmes were beyond the scope of my research. For example, I found that knowledge of Filipino celebrity gossip and plot points of soap operas were instrumental for me to gain the trust of respondents, especially lower-class audiences. Beginning our chats with 'did you see when...?' questions pertaining to popular talk shows and soap operas made respondents comfortable to disclose their thoughts and feelings, not only about television but also about other political and personal issues. Sharing common knowledge of popular culture created an atmosphere where participants felt they were not being judged as inferiors, especially as they knew that I was an academic (assumed to be critical of low culture) and perceived as middle or upper class (often presumed to be dismissive or disdainful of local television). Fieldwork involved not only interviews and participant observation among audiences in their everyday lives then. It also involved heavy viewing of television, taking copious notes about program content, collecting video material from ABS-CBN and GMA and analysing video material used during interviews. Fieldwork then was not solely focused on data-gathering [it involved a cyclical process of doing analysis (of audience interpretations and video materials), following new leads, dropping old ones and visiting and revisiting respondents.

NOTES

Introduction: The Poverty of Television

1 Speaking 25 years ago, David (2001[1976], 42) remarked that Philippine social science 'may probably be labelled as the sociology of coping mechanisms' in its preoccupation with detailed records of 'the way of life' of slum communities without sufficient analysis of wider social structures that produce and reproduce inequality, poverty and deprivation.

1. The Moral Turn: From First Principles to Lay Moralities

1 Silverstone's *Media and Morality* makes an equivalence of 'morality', 'the moral' and 'the ethical' and poses these three terms against 'ethics'. Whilst morality/moral/ethical are the general first principles of the good, ethics is 'the application of those principles in particular social or historical, personal or professional contexts' (Silverstone 2007, 7). This definitional distinction is useful for media studies especially, as there are frequent references to 'codes of ethics', particularly of newsrooms and media organizations. But as Couldry (2008a) points out, in other disciplines the distinction is often reversed, as ethics pertains to universal codes, and morality pertains to local or applied moral codes. In a review of the anthropology of moralities literature, Heintz (2009) makes similar observations as Couldry and admits that the words are used interchangeably by authors. Heintz defines morality as a 'set of principles and judgments based on cultural concepts and beliefs by which humans determine whether given actions are right or wrong' (3). I adopt Heintz's and Zelizer's (2008) definition of morality when I discuss ordinary people's judgments of right and wrong in relation to suffering and its mediation.

2 It must be noted that other media ethics scholars propose or emphasize different moral principles. See Dayan (2007) for a critique of Silverstone's lack of engagement with justice as an important principle that should guide media work. Meanwhile, Couldry (2006, 2008a) proposes a more practice-based virtue ethics approach that reflects on notions of the good rather than the imposition maximalist obligations. But, citing Kantian scholars such as O'Neill (1996), Couldry (2008a, 6) also admits that both deontological and virtue approaches share important similarities. Indeed, though they may be informed by different philosophical traditions, Couldry (2006, 2008a) and Silverstone (2007) are both able to pass judgment on the ethics of particular practices of media representation and audience activity.

3 More than 90 per cent of the Philippine people are Christian, with a majority being Roman Catholic. Around 8 per cent are Muslims, and 2 per cent have no religious affiliation. Catholicism was introduced by Spain, as their project of colonization went hand in hand with evangelization from 1521 to 1898. See Rafael (1988) for an historical analysis of processes by which colonial ideas and practices were adopted, contested and contracted by the Tagalogs in line with precolonial beliefs and practices.

2. Theorizing Mediated Suffering: Ethics of Media Texts, Audiences and Ecologies

1 In November 2013, ABS-CBN launched the ABS-CBN Mobile (the SIM pack costs 30 pesos), which allows its subscribers not only to text, to call and to surf the Internet but also to download a mobile application (iWanTV app) that allows them to watch TV programs of ABS-CBN, ANC, Studio 23, Nickelodeon, E! and Knowledge Channel in their phones, tablets or computers. At the same time, it also allows them to interact with and get exclusive information, photos and videos of ABS-CBN artists. On its launching, ABS-CBN gave away 100,000 SIM cards to Typhoon Haiyan survivors to help them communicate with their relatives (Lastrilla 2013).
2 Funding of terrestrial channels is fully advertiser-dependent. There are no subscription fees or licenses for publicly available ('non-cable') channels. The state-owned channels are subsidized by the government but have limited production budgets and consequently suffer from low viewership.
3 The Philippines is known as the 'texting capital of the world', with an average of 600 text messages sent by an individual user each month (Dimacali 2010).
4 Of the 70+ cable television channels, around 12 carry Philippine-produced and -oriented content (the most popular of which are the English language ABS-CBN News Channel, DZMM Teleradyo and Cinema One). The rest carry international programming, such as CNN, HBO, ESPN, StarTV, Disney Channel and MTV.
5 In industry parlance, the term 'local' (often used as counterpoint to 'foreign') refers to media content consumed by the national Filipino audience. Area- or province-specific television content is rare and often confined only to local evening newscasts in key markets such as Cebu, Davao, Dagupan and Iloilo.

3. Audience Ethics: Mediating Suffering in Everyday Life

1 The use of 'lower class' (rather than 'working class') is consistent with previous academic scholarship (Aguilar 2003, 2005a; Benedicto 2009; Davis and Hollnsteiner 1969; Doronila 1985; Guevara et al. 2009; Hollnsteiner 1973; Tadiar 2004; Tolentino 2007) and industry research (AGB Nielsen 2010a; McCann Erickson 2009) on class in the Philippines. The term 'lower class' in the Philippine context includes many of the unemployed and irregularly employed, as the country has a total unemployment rate of 27.5 per cent (SWS 2014). The more established Euro-American term 'working class' in this context may then be misleading, for the term 'working' connotes employment. This book uses the term 'lower class' without intending any negative value judgment on low-income individuals.
2 The use of five SEC categories (AB, C1, C2, D and E) was established in the Philippines in 1991. But in 2010, aiming to put forward a more 'inclusive approach in putting the

best possible SEC system', the NSO together with the University of the Philippines and Marketing and Opinion Research Society (MORES) proposed the 1SEC that used numbers instead of letters for the household groupings. The least spending household is grouped under Cluster 1 while the highest spending household is under Cluster 9. With 1SEC, about 45 per cent of the Philippine population falls under the least spending household, while 35 and 20 per cent belongs to the moderate and highest spending households respectively. These categories are nonetheless still being 'sharpened' and worked out (Bersales 2013).

3 Friendster is a social networking site that preceded Facebook. In 2010, it claims 100 million registered users, and 30 million monthly users, mainly in Southeast Asia (Miller and Madianou 2010). During the time of fieldwork, upper-class respondents have migrated to Facebook from Friendster, but many of my middle- and lower-class respondents maintained Friendster accounts at that time. Due to technological problems and loss of market shares, Friendster shifted away from a social networking site to a gaming platform in 2011 (Fiegerman 2014).

4 I am unable to explain how and why these specific neighbouring communities developed different affinities for the networks. I suspect, however, that the common practice (especially among women) of discussing television programmes and celebrity gossip in the basketball court area of Park 7 encourages loyal and devoted viewing of a common channel. The phrase I frequently heard there – *huli ka na sa balita'* (you're late to the news) – is used as a sanction for failing to keep up with community gossip and narratives in soaps, talk shows and the news.

5 I observed that channel-switching happens more frequently with news rather than entertainment programmes. There is a common public perception, especially magnified during the fieldwork period which preceded the national elections, that ABS-CBN is more biased than GMA. ABS-CBN news anchors, journalists and celebrities have widely known personal ties to politicians, including the famous broadcast journalist Korina Sanchez, who was fiancée to then-presidential candidate Mar Roxas. Their October 2009 wedding was marketed as the 'wedding of the year' and was broadcasted by ABS-CBN.

6 *Sari-sari* stores, literally 'variety stores', are privately owned convenience stores located in most low-income and middle-class streets and neighbourhoods. These stores typically sell the sachet sizes of household commodities. These are often places where neighbours meet and gossip or even watch television together on small black-and-white screens.

7 ETC is a Filipino-owned cable channel that airs primarily American imported television programmes such as *Project Runway* and *TMZ* and TV series such as *Glee, New Girl* and *Friends*. Though it produces local content, its programmes are all in English and skew to an upper-class demographic. AXN is a Singapore-based cable channel owned by Sony Pictures Entertainment that carries mostly American programming (e.g., *The Amazing Race*) and produces original lifestyle and talk show programmes that target English-speaking viewers in Southeast Asia.

8 The May 2009 sex video scandal involved the leak of a video of a well-known male plastic surgeon and a famous actress dancing in their underwear in front of a bedroom mirror – without the actress' knowledge. The video clip was uploaded anonymously on YouTube (allegedly stolen from the plastic surgeon's computer) and caused a scandal involving widely publicized legal investigations, inquiries in the Philippine Senate (covered live by the news) and marathon coverage in talk shows.

9 Though the *Big Brother* franchise originated from the Dutch production company Endemol.

10 *Big Brother* in the US involves housemates nominating and voting off their other housemates. The UK version involves audience participation in 'evicting one of the nominated housemates. Meanwhile, *Pinoy Big Brother* compels audiences to save their favourite housemate among those nominated. Connivance and strategy are downplayed in the local version in line with cultural norms of conflict avoidance (Jocano 1997). The show emphasizes instead relationship-building, cooperation and the accomplishment of tasks/chores, as housemates seek approval from 'Big Brother' and the viewing public who vote every week to save their favourites.

11 Brewis and Jack (2010, 257) however identify that whilst the typical chav is imagined to be mainly white, young and poor, chav can also refer to being brand-conscious and 'having a particular penchant for Burberry, Nike, Louis Vuitton and Adidas.' In contrast, the Filipino *jologs* has a closer association with a lack of money. Though *jologs* can be read of people who purchase the wrong kinds of clothes and accessories, often this is tied to their inability to purchase authentic versions of clothes and accessories, resorting instead to pirated, knock-off, or generic/'unbranded apparel.

12 The phrase 'word of hatred' I borrow from Skeggs (2005), who refers to chav as a 'word of hatred' for 'identifying objects of hatred, fear, anxiety and threat'.

13 Bench is a mass-market brand and is one of the most respected companies in the Philippines. Its continued success, especially in recent years, is explained by its continued popularity with its target audience as well as attracting the upper class through edgy marketing stunts, such as the biannual Bench Underwear Fashion Show.

14 Santiago (2010) astutely suggests that the public criticism towards the behaviour of *jologs* celebrity Marian Rivera is underpinned by middle- and upper-class standards of femininity.

15 *Jejemon* is the equivalent of *jologs*, but it pertains to their presence and practices in new media environments, such as in social networking sites. The recent popularization of Facebook among lower-class Filipinos has resulted in the coinage of the derogatory term *jejemon*, which refers to the particular and 'bad' use of English and text-speak in new media (see Ong and Cabañes 2011).

16 It is a common upper-class practice to speak in English or Taglish even in casual social interactions. Many upper-class Filipinos have difficulty speaking exclusively in Tagalog, and reading Tagalog texts. There is a correlation between class and the quality and frequent use of English.

4. Entertainment: Playing with Pity

1 Though *Wowowee* had been successful at being popular to many Filipinos and thus enjoyed high audience ratings, alongside its critics, *Wowowee* faced many problems, including the issue of the relationship of Willie Revillame (the programme's host) with ABS-CBN management that led for the show to end in 2010. Months after, when TV5 launched *Willing Willie*, the first show for Willie after he moved to TV5, ABS-CBN accused him and the network of copying *Wowowee's* format into the show. *Willing Willie* was eventually reformatted to *Willtime Bigtime*, partly in response to a public scandal (see Faustino 2011), but ended its run in January 2013. Revillame has been on mainstream media hiatus ever since, though actively involved in politicians' campaigns (ABS-CBN News Online 2013).

2 Jeepneys are public utility vehicles originally built from spare parts of American military jeeps from World War II. They are the most popular forms of transport in urban centres. Jeepney drivers were recently featured in a BBC2 documentary that featured the Philippines as the *Toughest Place to Be a [...] Bus Driver* (see Ropeta 2011).

3 He was quoted after the tragedy as having said, 'Is it a sin to give people hope?' (Isis International 2006).

4 Print media journalists expressed scathing criticism toward not just the lack of planning and crowd control of ABS-CBN Staff, but also the whole premise of *Wowowee* as a programme. However, television news coverage of the disaster was careful, to say the least. From my interviews with managers in both networks, ABS-CBN was naturally perceived to be too close to the situation to provide fair reporting. Meanwhile rival GMA practised self-censorship in being overtly careful so as not to appear that they were taking advantage of the tragedy for viewer affection.

5 One of the ABS-CBN producers was quoted, 'Even with all the dead around, many people were still asking for raffle tickets' (Doyo 2006).

6 Being selected as part of the ten 'Hep Hep Hooray' participants is based on luck, with odds of about 10 to 5,000, as the studio has a seating capacity of 5,000 (Isis International 2006). But after repeated viewings of *Wowowee*, one notices programme conventions of attempting to select a diverse mix of participants that often includes (1) predominantly older women and a few older men, (2) an attractive young woman or two, (3) a Filipino-American on holiday (seated in a section separate from general admission), (4) a well-dressed middle- or upper-class special guest and (5) a Caucasian or East Asian foreigner, presumably a guest of a Filipino in the audience.

7 Previous studies have identified that people with direct experiences of an event tend to be more critical of the media representation of the event (Madianou 2005b; Philo 1990). In this case, people who self-identified as sufferers were more engaged with the fates of contestants in the show, and drew from more expansive symbolic resources to judge participants.

8 First, contestants with darker skin are assumed by my respondents to be poorer than those with fairer skin. Certainly there is some element of 'colonial mentality' or 'white love' (Rafael 2000) here in assuming that wealth and comfort are properties of mixed-blood Filipino *mestizos* with Spanish, American and Chinese heritage (Aguilar 1998). However, this belief is also tied to assumptions that darker skin is a function of doing physical labour under the heat of the tropical sun and that fairer skin is a result of doing indoor office-based work. Second, contestants who are thin rather than fat are assumed to be poorer, presumably from a speculation on people's (lack of) access to food. Third, contestants who are older rather than younger tend to be cheered on by audiences more, as they are assumed to be less likely to be employed or employable. The fact that these contestants are on television rather than at work at twelve o'clock on a weekday signals that some are unemployed whilst others took time off from work. Fourth, contestants' clothing is also appraised for looking more or less expensive. People wearing rubber slippers and old or dirty clothes are judged to be more likely to be poor. Fifth, because contestants' voices are heard in the segment when they shout 'Hep Hep' or 'Hooray', audiences also evaluate their accents and manners of speaking. Contestants whose accents betray their provincial origins (e.g., people from Visayan provinces pronounce 'hep hep' as 'heep heep') are assumed to be in greater need than Manila-based or straight-English-speaking contestants. Provincial contestants are assumed to have traveled/migrated to Manila in search of jobs, opportunities and social services in the mediated centre absent in their hometown. Sixth, any contestant who looks and sounds like they are non-native-Filipinos – either foreigners or Filipino-Americans – are automatically perceived to be not poor. Hosts and audiences alike treat these contestants as novelty characters playing for fun rather than actively seeking money.

9 Aguilar (1998) has once remarked about Filipinos regarding life as a 'game of chance'. In many ways, *Wowowee* is the televisual equivalent of Filipino gambling games such as cockfighting, poker, spider-fighting and boxing previously studied by anthropologists (Aguilar 1998; Hollnsteiner 1973; Jocano 1975).

10 In 2007, *Wowowee* underwent investigation from the Department of Trade and Industry. In one segment, a prop is used to reveal numbers containing the amount of money the contestant would win. On live TV, the prop malfunctioned and revealed more numbers than it should. This seemed to suggest that the programme host has total discretion to pull out different numbers to award different cash amounts based on the contestant at hand–a practice that violates standard practices of game shows and raffles (GMA News Online 2010).

11 A euphemism for sex worker.

12 Here I recall Cannell's (1999) meditation on the artful strategy of conveying reluctance as one way in which the powerless gain the attention and sympathy of the powerful in Filipino social life.

5. News: Recognizing Calls to Action

1 There are more than 10 million Filipinos overseas workers, about 11 per cent of the total population (CFO 2012).

2 Based on these comparisons, one can speculate that there is greater likelihood for agency to be depicted in the case of Chinese case than that of Burma, given journalists' acknowledgment of the Chinese government's efficient response.

REFERENCES

Abrera, E. 1999. 'The Appliance of Our Lives'. In *Pinoy Television: The Story of ABS-CBN – The Medium of Our Lives.*, edited by T. Sioson-San Juan. Pasig City: ABS-CBN Publishing.

ABS-CBN News Online. 2010a. 'Catholic Bishops Want Condom Ads Banned'. ABS-CBN News Online, 3 March. Online: http://www.abs-cbnnews.com/nation/03/03/10/catholic-bishops-want-condom-ads-banned (accessed 2 December 2010).

_____. 2010b. 'Filipino Internet Users Most Engaged in Social Media: Survey'. ABS-CBN News Online, 8 April. Online: http://www.abs-cbnnews.com/lifestyle/04/08/10/filipino-internet-users-most-engaged-social-media-survey (accessed 25 February 2014).

_____. 2013. 'Willie Revillame in tears as show goes off air'. ABS-CBNnews. *October 12.* Online: http://www.abs-cbnnews.com/entertainment/10/12/13/willie-revillame-tears-show-goes-air (accessed 1 April 2014).

Abu-Lughod, L. 2002. 'Egyptian Melodrama—Technology of the Modern Subject?'. In *Media Worlds*, edited by F. Ginsburg, L. Abu-Lughod and B. Larkin. Berkeley and Los Angeles: University of California Press.

AGB Nielsen. 2006. *AGB Nielsen Establishment Survey*. Manila: AGB Nielsen.

_____. 2009. *AGB Nielsen Quarter 1 2009 Mega Manila Household Ratings Survey.*

_____. 2010. *The Philippine Media Landscape*. Manila: AGB Nielsen. Online: http://www.slideshare.net/bingkimpo/one-nielsen-press-briefing-28-march-2011 (accessed 2 April 2014).

Aguilar, F. V. 1998. *Clash of the Spirits: The History of Power and Sugar Planter Hegemony on a Visayan Island*. Honolulu: University of Hawaii Press.

_____. 2003. 'Global Migrations, Old Forms of Labor, and Transborder Class Relations'. *Southeast Asian Studies* 41, no. 2: 137–61.

_____. 2005. 'Tracing Origins: Ilustrado Nationalism and the Racial Science of Migration Waves'. *The Journal of Asian Studies* 64, no. 3: 605–37.

Albert, J. 2013. 'Big Data: Big Threat or Big Opportunity for Official Statistics?' 18 October. Online: http://www.nscb.gov.ph/statfocus/2013/SF_102013_OSG_bigData.asp (accessed 1 April 2014).

Alcock, P. 2006. *Understanding Poverty*, 3rd ed. Hampshire: Palgrave Macmillan.

Altheide, D. and P. Snow. 1979. *Media Logic*. Beverly Hills, CA: Sage.

Anderson, B. 1983. *Imagined Communities: Reflections on the Origins and Spread of Nationalism*. London: Verso.

Ang, I. 1996. 'Ethnography and Radical Contextualism in Audience Studies'. In *The Audience and Its Landscape*, edited by J. Hay, L. Grossberg, and E. Wartella, 247–62. Boulder, CO: Westview.

Antipinoy.com. 2010. 'Dancing Girls on TV: Cultural Dehumanization'. *Antipinoy.com*, 27 October. Online: http://antipinoy.com/dancing-girls-on-tv (accessed 2 December 2010).

Appiah, K. A. 2006. *Cosmopolitanism: Ethics in a World of Strangers*. New York: W.W. Norton & Co.

Arceo, T. 2004. 'Big Things Come in Small Packages'. *Change/Agent*, June. Online: http://www.synovate.com/changeagent/index.php/site/full_story/big_things_come_in_small_packages (accessed 2 December 2010).

Archetti, E. P. 1997. 'The Moralities of Argentinean Football'. In *The Ethnography of Moralities*, edited by S. Howell. London: Routledge.

Arendt, H. 1963. *On Revolution*. New York: Viking.

———. 1968. *Men in Dark Times*. New York: Harcourt Brace & World.

Arneson, R. J. 2004. 'Moral Limits on the Demands of Beneficence?'. In *The Ethics of Assistance: Morality and the Distant Needy*, edited by D. K. Chaterjee. Cambridge: Cambridge University Press.

Ashuri, T. and A. Pinchevski. 2009. 'Witnessing as a Field'. In *Media Witnessing: Testimony in the Age of Mass Communication*, edited by P. Frosh and A. Pinchevski, 133–57. London: Palgrave.

Atillo, A. B. and J. Serrano. 2004. 'The Face of Altruism among Culturally Diverse Low Income Households and Individuals in Davao del Sur'. In *Beyond the Household: Giving and Volunteering in Six Areas in the Philippines*, edited by R. L. Fernan III. Quezon City: NCPAG University of the Philippines.

Attridge, D. 2004. 'Ethical Modernism: Servants and Others in J. M. Coetzee's Early Fiction'. *Poetics Today* 25, no. 4: 653–71.

Bankoff, G. 2003. *Cultures of Disaster: Society and Natural Hazards in the Philippines*. London: Taylor & Francis.

———. 2007. 'Dangers to Going It Alone: Social Capital and the Origins of Community Resilience in the Philippines'. *Continuity and Change* 22, no. 2: 327–55.

Baudrillard, J. 1994. *The Gulf War Did Not Take Place*. Sydney: Powerful Publications.

———. 2001. 'The Mind of Terrorism', *Le Monde*, 2 November.

Bauman, Z. 2001. 'Whatever Happened to Compassion?' In *The Moral Universe*, edited by T. Bentley and I. Hargreaves. London: Demos.

BBC News Online. 2001. 'Dyke: BBC Is "Hideously White"'. BBC News Online, 6 January, http://news.bbc.co.uk/1/hi/scotland/1104305.stm (accessed 2 December 2010).

Bello, W. 1999. *Dark Victory: The United States and Global Poverty*, 2nd ed. London: Pluto Press.

Benedicto, B. 2009. 'Shared Spaces of Transnational Transit: Filipino Gay Tourists, Labour Migrants, and the Borders of Class Difference'. *Asian Studies Review*, 33: 289–301.

Bird, E. 1998. *For Enquiring Minds: A Cultural Study of Supermarket Tabloids*. Knoxville: University of Tennessee Press.

Boltanski, L. 1999. *Distant Suffering: Morality, Media and Politics*. Cambridge: Cambridge University Press.

Bordadora, N. 2011. 'Palace Exec Blames Media for Dip in Aquino's Ratings'. *Inquirer.net*, 2 April. Online: http://newsinfo.inquirer.net/breakingnews/nation/view/20110402-329004/Palace-exec-blamesmedia-for-dip-in-Aquinos-ratings (accessed 5 May 2011).

Born, G. 2008. *The Normative, Institutional and Practical Ethics: For an Anthropological Ethics of Media*. University of Cambridge: Ethics of Media Conference.

Born, G. and T. Prosser. 2001.'"Culture and Consumerism: Citizenship, Public Service Broadcasting and the BBC's Fair Trade Obligation'. *Modern Law Review*, 64, no. 5: 657–87.

Bourdieu, P. 1986[1979]. *Distinction: A Social Critique of the Judgment of Taste*. Translated by R. Nice. Cambridge, MA: Harvard University Press.

Brand, R. 2009. 'Witnessing Trauma on Film'. In *Media Witnessing: Testimony in the Age of Mass Communication*, edited by P. Frosh and A. Pinchevski, 198–214. London: Palgrave.

Brewis, J. and G. Jack. 2010. 'The Ambiguous Politics of Gay Chavinism'. *Sociology*, 44, no. 2: 251–68.

Buckingham, D. 2000. *The Making of Citizens: Young People, News and Politics*. London: Routledge.

Burgos, A. 2010. 'TV a Less Watched King, Survey Says'. ABS-CBN News Online, 16 October. Online: http://www.abs-cbnnews.com/nation/10/16/10/tv-less-watched-king-survey-says (accessed 2 December 2010).

Business Mirror. 2013. 'Cell Phones Bolster PHL's Microfinance Growth'. *Business Mirror*, 12 January. Online: http://businessmirror.com.ph/index.php/en/news/top-news/7588-cell-phones-bolster-phl-s-microfinance-growth (accessed 1 April 2014).

Butler, J. 2004. *Precarious Life: Powers of Mourning and Violence*. London: Verso.

_____. 2009. *Frames of War: When Is Life Grievable?* London: Verso.

Cabañes, J. 2011. 'The Crisis of Voice in Multicultural, Mediated Manila'. Paper Presented at the ICA Annual Conference, Boston, USA.

_____. 2013. 'The Mediation of Multiculturalism in Manila: Participatory Photography with Indians and Koreans in Manila'. PhD diss., University of Leeds.

Cannell, F. 1999. *Power and Intimacy in the Christian Philippines*. Cambridge: Cambridge University Press.

Carat Philippines. 2009. *2009 Philippine Media Landscape*. Manila: Philippines.

Center for Media Freedom and Responsibility. 2007. *Philippine Press Freedom Report 2007*. Manila: CMFR.

Centre for Research on Epidemiology and Disaster. 2010. *Annual Disaster Statistical Review 2009: The Numbers and Trends*. Online: http://cred.be/sites/default/files/ADSR_2009.pdf (accessed 2 December 2010).

Chouliaraki, L. 2006. *The Spectatorship of Suffering*. London: Sage.

_____. 2010 'Post-humanitarianism: Humanitarian Communication Beyond a Politics of Pity'. *International Journal of Cultural Studies*, 13, no. 2: 107–26.

_____. 2012. 'Re-mediation, Inter-mediation, Trans-mediation.' Journalism Studies 14(2): 267–83.

_____. 2013. *The Ironic Spectator: Solidarity in the Age of Post-humanitarianism*. Cambridge: Polity.

Claudio, L. 2010. 'Memories of the Anti-Marcos Movement: The Left and the Mnemonic Dynamics of the Post-Authoritarian Philippines'. *South East Asia Research* 18, no. 1: 33–66.

Clement, C. 1996. *Care, Autonomy, and Justice: Feminism and the Ethic of Care*. Oxford: Westview Press.

CNN International. 2005. 'World Must "Wake Up" to Disasters'. CNN International, 4 January. Online: http://edition.cnn.com/2005/WORLD/asiapcf/01/03/un.egeland.disasters (accessed 10 December 2010).

Cohen, S. 2001. *States of Denial: Knowing about Atrocities and Suffering*. London: Polity.

Commission on Filipino Overseas. 2012. *Stock Estimate of Overseas Filipinos*. Online: http://www.cfo.gov.ph/images/stories/pdf/StockEstimate2012.pdf (accessed 2 April 2014).

Comte-Sponville, A. 2001. *A Small Treatise on the Great Virtues*. New York: Henry Holt and Co.

Constantino, R. 1985. *Synthetic Culture and Development*. Quezon City: Foundation for Nationalist Studies.

Contreras, V. 2011. 'Don't Dumb Down Viewers, TV Urged'. *Inquirer.net*, 11 April. Online: http://newsinfo.inquirer.net/inquirerheadlines/nation/view/20110411-330486/ Dont-dumb-downviewers-TV-urged (accessed 5 May 2011).

Cottle, S. 2006. *Mediatized Conflict*. Basingstoke: Open University Press.

Cordova, J. 2011. 'Why the Filipino Elite Revile Willie Revillame'. *The Asian Correspondent*, 16 April. Online: http://asiancorrespondent.com/52569/why-the-filipino-elite-revile-willie-revillame (accessed 5 May 2011).

Cornelio, J. 2011. 'Religious Identity and the Isolated Generation: What Being Catholic Means to Religiously Involved Filipino Students Today'. PhD diss., National University of Singapore. Singapore.

Coronel, S. 1999. 'Lords of the Press'. In *From Loren to Marimar: The Philippine Media in the 1990s*, edited by S. Coronel. Quezon City: Philippine Center for Investigative Journalism
_____. 2001. 'The Media, the Market and Democracy: The Case of the Philippines'. *The Public*, 8: 109–26.
_____. 2006a. '*Wowowee*: Television and the Perils of Peddling Dreams'. *The Daily PCIJ*, 9 February. Online: http://www.pcij.org/blog/?p=593 (accessed 2 December 2010).
_____. 2006b. '*Wowowee* and the Women of 200 P De la Cruz Street'. *Philippine Center of Investigative Journalism*, 1 March. Online: http://pcij.org/stories/wowowee-and-the-women-of-200-pde-lacruz-st (accessed 2 December 2010).
_____. 2006c. '*Wowowee*: A Filipino Tragedy'. *The Daily PCIJ*, February 4. Online: http://www.pcij.org/blog/?p=584 (accessed 2 December 2010).

Couldry, N. 2000. *The Place of Media Power: Pilgrims and Witnesses of the Media Age*. London: Routledge.
_____. 2003. *Media Rituals: A Critical Approach*. London: Routledge.
_____. 2006. *Listening Beyond the Echoes: Media, Ethics and Agency in an Uncertain World*. Boulder, CO: Paradigm Books.
_____. 2008a. 'Communicative Virtue and the Construction of a Global Media Ethics'. Paper presented at the Ethics of Media Conference. CRASSH, University of Cambridge, Cambridge, UK.
_____. 2008b. 'Mediatization or Mediation? Alternative Understandings of the Emergent Space of Digital Storytelling'. *New Media & Society* 10: 373–91.
_____. 2009. 'My Media Studies: Thoughts from Nick Couldry'. *Television and New Media* 10: 40–2.
_____. 2010. *Why Voice Matters: Culture and Politics after Neoliberalism*. London: Sage.
_____. 2012. *Media Society World: Social Theory in a Digital Age*. Cambridge: Polity.

Couldry, N., M. Madianou and A. Pinchevski (eds). 2013. *Ethics of Media*. London: Palgrave Macmillan.

Couldry, N., S. Livingstone and T. Markham. 2007. *Media Consumption and Public Engagement: Beyond the Presumption of Attention*. London: Palgrave Macmillan.

Cruz, M. 2007. '*Wowowee* Suspended'. *Inquirer.net*, 7 June. Online: http://showbizandstyle. inquirer.net/breakingnews/breakingnews/view/20070607-69911/'Wowowee'_ suspended (accessed 10 December 2010).

Cruz, N. 2009. '"Ondoy" Was Great Equalizer'. *Inquirer.net*, 30 September. Online: http:// opinion.inquirer.net/inquireropinion/columns/view/20090930-227604/Ondoy-was-greatequalizer (accessed 30 September 2010).

Cruz, N. H. 2006. 'Stampede Shows How Desperately Poor Pinoys Are'. *Inquirer.net*, 5 February. Online: http://www.inquirer.net/specialfeatures/ultrastampede/view.php? db=0&article=20060205-65233 (accessed 2 December 2010).

Curato, N. 2013. '#MillionPeopleMarch and the Limits of Playful Citizenship'. *Rappler,* 27 August. Online: http://www.rappler.com/thought-leaders/37401-millionpeople march-citizenship-social-media (accessed 27 August 2013).

Cutchin, M. P. 2002. 'Ethics and Geography: Continuity and Emerging Syntheses'. *Progress in Human Geography* 26, no. 5: 656–64.

Dalton, S., H. Maddon, K. Chamberlain, S. Carr and A. C. Lyons. 2008. '"It's Gotten a Bit Old, Charity": Young Adults in New Zealand Talk about Poverty, Charitable Giving and Aid Appeals'. *Journal of Community & Applied Social Psychology* 18: 492–504.

Danahar, P. 2008. 'Burma and China: Tale of Two Disasters'. BBC News Online, 19 May, http://news.bbc.co.uk/1/hi/world/asia-pacific/7407927.stm (accessed 2 December 2010).

Dancel, R. 2010. 'Dear Willie'. *I Flip Daily*, 6 May. Online: http://pininggapura.wordpress. com/2010/05/06/dear-willie (accessed 2 December 2010).

Das, V. 1995. *Critical Events*. Oxford: Oxford University Press.

David, R. S. 2001a[1976]. 'The Sociology of Poverty or the Poverty of Sociology? A Brief Note on Urban Poverty Research'. In *Reflections on Sociology & Philippine Society*, by R. S. David. Quezon City: University of the Philippines Press.

_____. 2001b[1982]. 'Sociology and Development Studies in the Philippines'. In *Reflections on Sociology & Philippine Society*, by R. S. David. Quezon City: University of the Philippines Press.

_____. 2001c[2000]. 'Public Service Broadcasting: A Talent to Amuse or a Mission to Explain?' In *Reflections on Sociology & Philippine Society*, by R. S. David. Quezon City: University of the Philippines Press.

Davis, W. G. and M. Hollnsteiner. 1969. 'Some Recent Trends in Philippine Social Anthropology'. *Anthropologica* 11, no. 1: 59–84.

Dayan, D. 2007. 'On Morality, Distance and the Other: Roger Silverstone's *Media and Morality.*" *International Journal of Communication*, 113–22.

Dayan, D. and E. Katz. 1992. *Media Events: The Live Broadcasting of History*. Cambridge, MA: Harvard University Press.

De Jesus, M. Q. 2011. 'TV, *Willing Willie*, the Public Sphere' in *Rex Crisostomo's Blog*, 26 April. Online: http://rexcrisostomo.blogspot.com/2011/04/tv-willing-willie-public-sphere.html (accessed 5 May 2011).

De Quiros, C. 2006. 'Tragedy'. *Inquirer.net*, 5 February. Online: http://www.inquirer.net/ specialfeatures/ultrastampede/view.php?db=0&article=20060205-65235 (accessed 2 December 2010).

_____. 2009a. 'Three'. *Inquirer.net*, 30 September. Online: http://opinion.inquirer. net/inquireropinion/columns/view/20090930-227605/Three (accessed 30 September 2009).

_____. 2009b. '"Filipino Resilience"'. *Inquirer.net*, 20 October. Online: http://opinion. inquirer.net/inquireropinion/columns/view/20091020-231079/Filipino-resilience/ (accessed 2 December 2010).

De Vera, R. 2009. 'Epic Reincarnation'. *Inquirer.net*, 19 December. Online: http://lifestyle. inquirer.net/super/super/view/20091219-242785/Epic-Reincarnation (accessed 2 December 2010).

De Zengotita, T. 2005. *Mediated: How the Media Shapes Your World and the Way You Live in It*. New York: Bloomsbury.

Del Mundo, C. 1993. *Telebisyon: An Essay on Philippine Television*. Manila: Cultural Center of the Philippines.

Devilles, G. 2008. 'The Pornography of Poverty in *Serbis* and *Tribu*'. Paper Presented at the 2008 Southeast Asian Cinema Conference. Ateneo de Manila University, Quezon City, Philippines.

Dimacali, T. 2010. 'Philippines Still Text Messaging Champ – US Study'. GMA News Online, 18 February. Online: http://www.gmanews.tv/story/198832/philippines-still-text-messaging-champ-us-study (accessed 10 December 2010).

Dinglasan, R. 2013. 'Filipino Middle Class Drives Spending'. GMA News Online, 23 July. Online: http://www.gmanetwork.com/news/story/318755/economy/business/filipino-middle-class-drives-spending-nielsen-global-survey (accessed 1 April 2014).

Donnar, G. 2009. 'Passive Engagement and "The Face": The Possibility of Witnessing, Recognizing and Recovering Mediated Bodies in Suffering'. *Critical Perspectives on Communication, Cultural & Policy Studies*, 28, no. 2: 43–50.

Doronila, A. 1985. 'The Transformation of Patron-Client Relations and Its Political Consequences in Postwar Philippines'. *Journal of Southeast Asian Studies* 16, no. 1: 99–116.

Doyo, M. C. P. 2006. '*Wowowee* Pied Piper-ed the Poor'. *Inquirer.net*, 9 February. Online: http://www.inquirer.net/specialfeatures/ultrastampede/view.php?db=0&article=20060209-65561 (accessed 2 December 2010).

Drotner, K. 1992. 'Modernity and Media Panics'. In *Media Cultures: Reappraising Transnational Media*, edited by M. Skovmand and K. C. Schroder. London: Routledge.

Dy, A. 2009. 'New Reasons to Watch TV'. *Inquirer.net*, 17 October. Online: http://lifestyle.inquirer.net/2bu/2bu/view/20091017-231270/New_reasons_to_watch_TV (accessed 2 December 2010).

Ellis, J. 2000. *Seeing Things: Television in the Age of Uncertainty*. London: Tauris.

_____. 2009. 'Mundane Witnessing'. In *Media Witnessing: Testimony in the Age of Mass Communication*, edited by P. Frosh and A. Pinchevski. London: Palgrave Macmillan.

Espina, B. 2010. 'Filipino Compassion'. *Philippine Online Chronicles*, 28 August. Online: http://www.thepoc.net/commentaries/9497-filipino-compassion.html (accessed 2 December 2010).

Etzioni, A. 1995. *The Spirit of Community: Rights, Responsibilities and the Communitarian Agenda*. London: Fontana Press.

Evans, M. 2010. 'Exploring the Use of Twitter Around the World'. *Sysomos*, 14 January. Online: http://blog.sysomos.com/2010/01/14/exploring-the-use-of-twitter-around-the-world (accessed 2 December 2010).

Facebakers. 2010. 'Countries on Facebook – Facebook Statistics'. *Facebakers*, 10 December, http://www.facebakers.com/countries-with-facebook (accessed 10 January 2011).

Fassin, D. 2008 'Beyond Good and Evil? Questioning the Anthropological Discomfort with Morals'. *Anthropological Theory* 8, no. 4: 333–44.

Faustino, P. 'DSWD Call for Censure of Willie for Boy "Macho Dancer;" Willie Says Sorry'. GMA News Online, 28 March. Online: http://www.gmanetwork.com/news/story/216365/showbiz/dswd-calls-for-censure-of-willie-for-boy-macho-dancer-willie-says-sorry (accessed 1 April 2014).

Fiegerman, S. 2014. 'Friendster Founder Tells His Side of the Story, 10 Years After Facebook'. *Mashable*, February 3. Online: http://mashable.com/2014/02/03/jonathan-abrams-friendster-facebook/ (accessed 1 April 2014).

Flores, P. 2001. 'The Star Also Suffers: Screening Nora Aunor'. *Kasarinlan: Philippine Journal of Third World Studies* 16, no. 1.

Fraser, N. 1997. 'From Redistribution to Recognition? Dilemmas of Justice in a "Postsocialist" Age', in *Justice Interruptus: Critical Reflections on the "Postsocialist" Condition*, edited by N. Fraser. New York & London: Routledge.

Fraser, N. and A. Honneth. 2003. *Redistribution or Recognition?* London: Verso.

The Freeman. 2006. 'Probe Panel Summons ABS-CBN Execs: Security Blames Disaster on Mob'. 2006. *The Freeman*, 6 February. Online: http://www.pinoyexchange.com/forums/showthread.php?t=245687 (accessed 2 December 2010).

Frosh, P. 2007. 'Telling Presences: Witnessing, Mass Media, and the Imagined Lives of Strangers'. *Critical Studies in Media Communication* 23: 265–84.

Frosh, P. and A. Pinchevski. 2009. 'Introduction: Why Witnessing? Why Now?' In *Media Witnessing: Testimony in the Age of Mass Communication*, edited by P. Frosh and A. Pinchevski. Basingstoke: Palgrave Macmillan, pp. 1–22.

Galtung, J. and M. H. Ruge. 1965. 'The Structure of Foreign News: The Presentation of the Congo, Cuba, and Cyprus Crises in Four Norwegian Newspapers'. *Journal of Peace Research* 2: 64–91.

Garcia Canclini, N. 2001. *Consumers and Citizens: Globalization and Multicultural Conflicts.* Minneapolis: University of Minnesota Press.

Gatdula, D. 2014. 'Phl economy grows 7.2%, second fastest in Asia'. *The Philippine Star,* January 31. Online: http://www.philstar.com/headlines/2014/01/31/1285047/phl-economy-grows-7.2-second-fastest-asia (accessed 1 April 2014).

Gershon, I. 2010. *Breakup 2.0: Disconnecting over New Media.* Ithaca, NY: Cornell University Press.

Get Real Philippines. 2011. 'The Coming Fifth Anniversary of ABS-CBN's *Wowowee* Stampede Tragedy'. *Get Real Philippines*, 8 January. Online: http://getrealphilippines.blogspot.com/2011/01/coming-fifth-anniversary-of-abscbns.html (accessed 3 April 2011).

_____. 2013. 'HYPERLINK "http://getrealphilippines.com/blog/2013/11/cnns-anderson-cooper-versus-noynoy-aquino-on-th'HY*Get Real Philippines*, November 14. Online: http://getrealphilippines.com/blog/2013/11/cnns-anderson-cooper-versus-noynoy-aquino-on-the-typhoon-yolanda-situation-like-fact-versus-fiction/comment-page-29/ (accessed 15 November 2013).

Gibson, A. 1999. *Postmodernity, Ethics and the Novel: From Leavis to Levinas.* London: Routledge.

Giddens, A. 1991. *Modernity and Self Identity: Self and Society in the Late Modern Age.* Cambridge: Polity.

Gillespie, M. 1995. *Television, Ethnicity and Cultural Change.* London: Routledge.

Gilligan, C. 1982. *In a Different Voice: Psychological Theory and Women's Development.* Cambridge, MA: Harvard University Press.

Gilroy, P. 2004. *After Empire: Melancholia or Convivial Culture?* Abingdon: Routledge.

Ginzburg, C. 1994. 'Killing of a Chinese Mandarin: The Moral Implications of Distance'. *New Left Review* 208: 107–20.

GMA News Online. 2008. 'Bill Seeks to Penalize Illegal Cable TV, Internet Connections'. GMA News Online, 18 June. Online: http://www.gmanews.tv/story/101777/Bill-seeks-to-penalize-illegal-cable-TV-Internetconnections (accessed 2 December 2010).

_____. 2010a. 'DTI Sues ABS-CBN over *Wowowee* Game Mess'. GMA News Online, 15 January. Online: http://www.gmanews.tv/story/76658/DTI-sues-ABS-CBN-over-Wowowee-game-mess (accessed 2 December 2010).

_____. 2010b. 'GMA Network Bags Reader's Digest Most Trusted Brands Award'. 2010. GMA News Online, 5 May. Online: http://www.gmanews.tv/story/190145/gma-network-bags-readers-digest-most-trusted-brandsaward (accessed 2 December 2010).

GMA Research Department. 2004. *Cebu Audience Study.* Quezon City: Philippines.

Godinez, B. 2007. 'MTRCB Heads that Said "Cut!"' *Philippine Entertainment Portal*, 21 September. Online: http://www.pep.ph/features/looking-back/13859/mtrcb-heads-that-said-quotcutquot (accessed 2 December 2010).

Gregg, M. (2011) *Work's Intimacy.* Cambridge: Polity.

Guevara, M. C. C., A. M. Mayor and M. Racelis. 2009. *The Philippine Pilot Study of the Child Friendly Community Participatory Assessment Tools.* Quezon City: Childwatch International Research Network.

Habermas, J. 1989. *The Structural Transformation of the Public Sphere: An Inquiry into a Category of Bourgeois Society.* Translated by T. Burger with F. Lawrence. Cambridge, MA: MIT Press.

Hammersley, M. and P. Atkinson. 1995 [1983]. *Ethnography: Principles in Practice.* New York: Tavistock.

Heintz, M. 2009. 'Introduction: Why There Should Be an Anthropology of Moralities'. In *The Anthropology of Moralities,* edited by M. Heintz. New York, NY: Berghahn Books.

Hermitanio, M. 2009. 'Censorship Allegations Cause Friction Between Indie Cinema and MTRCB'. *Philippine Online Chronicles,* 24 June. Online: http://www.thepoc.net/thepoc-features/metakritiko/metakritiko-features/1591-censorship-allegations-cause-friction-between-indie-cinema-and-mtrcbpart-1-of-2.html (accessed 2 December 2010).

Hesmondhalgh, D. and S. Baker. 2010. *Creative Labour: Media Work in Three Cultural Industries.* Abingdon & New York: Routledge.

Hill, A. 2005. *Reality TV: Audiences and Popular Factual Television.* London: Routledge.

Ho, A. 2011. 'Philippines leads in income inequality in Asean, says study'. *Philippine Daily Inquirer,* 22 July. Online: http://business.inquirer.net/8377/philippines-leads-in-income-inequality-in-asean-says-study (accessed 23 June 2013).

Hochschild, A. 1983. *The Managed Heart: Commercialization of Human Feeling.* Berkeley & Los Angeles: University of California Press.

Hofileña. C. 2004. *News for Sale.* Quezon City: Philippine Center for Investigative Journalism.

Höijer, B. 2004. 'The Discourse of Global Compassion: The Audience and Media Reporting of Human Suffering'. *Media Culture & Society* 26: 513–31.

Hollnsteiner, M. 1973. 'Reciprocity in the Lowland Philippines'. In *Four Readings in Philippine Values,* 3rd revised and enlarged edition, edited by F. Lynch and A. De Guzman. Quezon City: Ateneo de Manila University Press.

Howell, S. 1997. 'Introduction'. In *The Ethnography of Moralities,* edited by S. Howell. London: Routledge.

Ignatieff, M. 1998. *The Warrior's Honor: Ethnic War and the Modern Conscience.* London: Vintage.

Ileto, R. 1979. *Pasyon and Revolution: Popular Movements in the Philippines, 1840–1910.* Quezon City: Ateneo de Manila University Press.

Illouz, E. 2003. 'From the Lisbon Disaster to Oprah Winfrey: Suffering as Identity in the Era of Globalization'. In *Global America? The Cultural Consequences of Globalization,* edited by Beck, U., Sznaider, N., and R. Winter. Liverpool: Liverpool University Press.

Inquirer.net. 2006. 'Helping the Poor'. *Inquirer.net,* 7 February. Online: http://www.inquirer.net/specialfeatures/ultrastampede/view.php?db=0&article=20060207-65352 (accessed 2 December 2010).

Isis International. 2006. 'Game Shows Brings Death in the Philippines: Media Corporate Responsibility Questioned in the *Wowowee*'. *Isis International,* 3 April. Online: http://www.isiswomen.org/index.php?option=com_content&view=article&id=68:game-show-brings-death-in-the-philippines-mediacorporate-responsibility-questioned-in-the-wowowee&catid=21:cim&Itemid=230 (accessed 2 December 2010).

Jankelevitch, V. 1986. *Traite des Vertus,* vol 2. Paris: Champs-Flammarion.

Jensen, K. B. 1995. *The Social Semiotics of Mass Communication*. London: Sage.

Jimenez-David, R. 2006. 'They Were Invited'. *Inquirer.net*, 7 February. Online: http://www.inquirer.net/specialfeatures/ultrastampede/view.php?db=0&article=20060207-65347 (accessed 2 December 2010).

Jocano, F. L. 1975. *Slum as a Way of Life*. Quezon City: University of the Philippines Press.

_____. 1997. *Filipino Value System: A Cultural Definition*. Manila: Punlad Research House.

Johnson, M. 2010. 'Diasporic Dreams, Middle-Class Moralities and Migrant Domestic Workers Among Muslim Filipinos in Saudi Arabia'. *The Asia Pacific Journal of Anthropology* 11, nos 3-4: 428–48.

Katz, E. and T. Liebes. 2007 '"No More Peace!": How Disaster, Terror and War Have Upstaged Media Events'. *International Journal of Communication*, 157–66.

Kendall, B. 2008. 'Two Disasters, Contrasting Reactions'. BBC News Online, 13 May. Online: http://news.bbc.co.uk/1/hi/world/asia-pacific/7399004.stm (accessed 2 December 2010).

Kenney, J. S. and D. Clairmont. 2009. 'Using the Victim Role as Both Sword and Shield: The Interactional Dynamics of Restorative Justice Sessions'. *Journal of Contemporary Ethnography* 38, no. 3: 279–307.

Kerkvliet, B. J. T. 1990. *Everyday Politics in the Philippines*. Berkeley & Los Angeles: University of California Press.

Kim, Y. 2005. *Women, Television and Everyday Life in Korea: Journeys of Hope*. London & New York: Routledge.

_____. 2007. 'The Rising East Asian "Wave": Korean Media Go Global'. In *Media on the Move: Global Flow and Contra-Flow*, edited by D. Thussu. London & New York: Routledge.

Kinnick, K. N., D. M. Krugman and C. T. Cameron. 1996. 'Compassion Fatigue: Communication and Burnout Toward Social Problems'. *Journalism & Mass Communication Quarterly*, Autumn 1996: 687–707.

Klein, B. and C. Wardle. 2008. '"These Two Are Speaking Welsh on Channel 4!": Welsh Representations and Cultural Tensions on *Big Brother 7*'. *Television and New Media* 9, no. 6: 514–30.

Kleinman, A., V. Das and M. Lock. 1997. 'Introduction'. In *Social Suffering*, edited by A. Kleinman, V. Das and M. Lock. Berkeley: University of California Press.

Koh, E. L. 2006. 'The Culture of Poverty'. *Filipinojournal.com* 20, no. 5. Online: http://www.filipinojournal.com/v2/index.php?pagetype=read&article_num= 09182006014359&latest_issue=V20-N5 (accessed 2 December 2010).

Kyriakidou, M. 2005. '"Feeling the Pain of Others": Exploring Cosmopolitan Empathy in Relation to Distant Suffering'. In *Democracy, Journalism and Technology: New Developments in an Enlarged Europe*, edited by N. Carpentier, P. Pruulmann-Vengerfeldt, K. Nordenstreng, M. Hartmann, P. Vihalemm, B. Cammaerts, H. Nieminen, and T. Olsson. Tartu: Tartu University Press.

_____. 2008. 'Rethinking Media Events in the Context of a Global Public Sphere: Exploring the Audience of Global Disasters in Greece'. *Communications* 33: 273–91.

Laidlaw, J. 2002. 'For an Anthropology of Ethics and Freedom'. *Journal of the Royal Anthropological Institute* 8, no. 2: 311–22.

Lastrilla, G. 2013. First ABS-CBN Mobile Store Officially Opens. *ABS-CBN*, December 12. Online: http://www.abs-cbn.com/updates/PR-First-ABS-CBN-Mobile-store-officially-opens?scid=24427B32-4106-4E14-BDBF-9CABCAFB73D6 (accessed 1 April 2014).

Lindlof, T. R. 1995. *Qualitative Communication Research Methods*. Thousand Oaks, CA: Sage.

Livingstone, S. 1999. 'New Media, New Audiences?' *London: LLSE Research Online*. Online: http://eprints.lse.ac.uk/archive/00000391(accessed 1 April 2014).

———. 2005. 'On the Relation Between Audiences and Publics'. In *Audiences and Publics: When Cultural Engagement Matters for the Public Sphere*, edited by S. Livingstone. Bristol: Intellect.

———. 2009. 'On the Mediation of Everything: ICA Presidential Address 2008'. *Journal of Communication* 59, no. 1: 1–18.

Lundby, K. 2009. 'Media Logic: Looking for Social Interaction'. In *Mediatization*, edited by K. Lundby, 101–19. New York: Peter Lang.

Lunt, P. and P. Stenner. 2005. '*The Jerry Springer Show* as an Emotional Public Sphere'. *Media, Culture & Society* 27, no. 1: 59–81.

Lynch, F. 1973 'Social Acceptance Reconsidered'. In *Four Readings in Philippine Values*, 3rd revised and enlarged edition, edited by F. Lynch and A. De Guzman. Quezon City: Ateneo de Manila University Press.

Madianou, M. 2005a. 'The Elusive Public of Television News'. In *Audiences and Publics: When Cultural Engagement Matters for the Public Sphere*, edited by S. Livingstone. Bristol: Intellect.

———. 2005b. *Mediating the Nation: News, Audiences and the Politics of Identity*. London: UCL Pres

———. 2009. 'Living with News: Ethnography and News Consumption'. In *The Routledge Companion to News and Journalism Studies*, edited by S. Allan. Abingdon: Routledge.

———. 2011. 'News as Looking-Glass: Shame and the Symbolic Power of Mediation *International Journal of Cultural Studies*.

Madianou, M. and D. Miller. 2011. 'Mobile Phone Parenting? Reconfiguring Relationships Between Filipina Migrant Mothers and Their Left-Behind Children'. *New Media & Society* 13.

Manila Standard Today. 2013. 'PH is Social Networking Capital of the World'. *Manila Standard Today*, May 21. Online: http://manilastandardtoday.com/2013/05/21/ph-is-social-networking-capital-of-the-world/ (accessed 1 April 2014)

Mankekar, P. 1999. *Screening Culture, Viewing Politics: An Ethnography of Television, Womanhood, and Nation Postcolonial India*. Durham, NC: Duke University Press.

Martin, K. 2014. 'Fitch Keeps Investment Grade Rating for Phl'. *The Philippine Star*, March 2. Online: http://www.philstar.com/business/2014/03/26/1305116/fitch-keeps-investment-grade-rating-phl (accessed 1 April 2013).

Martinez-Belen. C. 2009. 'A Grand Kapuso Fans Day in Cebu'. *Manila Bulletin*, 3 July. Online: http://www.mb.com.ph/articles/209012/a-grand-kapuso-fans-day-cebu (accessed 2 December 2010).

Maslog, C. C. 1990. *Philippine Mass Communication: A Mini-History*. Quezon City: New Day Publishers.

Mauss, M. 1966. *The Gift: Forms and Functions of Exchange in Archaic Societies*. Translated by Cunnison. London: Cohen & West.

Mawis, S. D. 2008. 'Katipunan's Young Breadwinners'. *The GUIDON*, 18 December. Online: http://inquiry.theguidon.com/2008/12/katipunan's-young-breadwinners (accessed 2 December 2010).

Mayer, V. 2011. 'Moral Tactics in a Film Production Economy'. Paper presented at the Moral Economies of Creative Labour Conference. Institute of Communication Studies, University of Leeds, UK.

McCann Erickson Philippines. 2009. *2008 Philippine Media Landscape*. Makati City: McCann Erickson.

McCoy, A. 2009. 'Rent-Seeking Families and the Philippine State: A History of the Lopez Family'. In *An Anarchy of Families: State and Family in the Philippines*, edited by A. McCoy. Madison: University of Wisconsin Press.

McKay, D. 2009. 'Borrowed Lives: Faith and Debt in Filipino Migrant Networks, Hong Kong and London'. Paper presented at the Diasporic Encounters, Sacred Journeys Conference. Keele University, Keele, UK.

Melhuus, M. 1997. 'The Troubles of Virtue: Values of Violence and Suffering in a Mexican Context'. In *The Ethnography of Moralities*, edited by S. Howell. London: Routledge.

Meyer, T. 2002. *Media Democracy: How the Media Colonize Politics*. Cambridge: Polity.

Midgeley, M. 1998. 'The Problem of Humbug'. In *Media Ethics*, edited by M. Kieran. London: Routledge.

Miles, M. B. and A. M. Huberman. 1994. *Qualitative Data Analysis: An Expanded Sourcebook*, 2nd ed. London & Thousand Oaks, CA: Sage.

Miller, Daniel. 1992. '*The Young and the Restless* in Trinidad: A Case of the Local and the Global in Mass Consumption'. In *Consuming Technologies: Media and Information in Domestic Spaces*, edited by R. Silverstone and E. Hirsrch. London: Routledge.

Miller, D. and M. Madianou. 2010. 'Should You Accept a *Friends* Request from Your Mother? and Other Filipino Dilemmas'. Paper presented at the LSE Migration Studies Unit Seminar. London School of Economics, London, UK.

Miller, David. 1992. 'Distributive Justice: What the People Think'. *Ethics* 102, no. 3: 555–93.

Moeller, S. 1999. *Compassion Fatigue: How the Media Sell Disease, Famine, War and Death*. London & New York: Routledge.

_____. 2002. 'A Hierarchy of Innocence: The Media's Use of Children in the Telling of International News'. *Press/Politics* 7, no. 1: 36–56.

Moorti, S. 1998. 'Cathartic Confessions or Emancipatory Texts? Rape Narratives on *The Oprah Winfrey Show*'. *Social Text* 57: 83–102.

Morley, D. 2006. 'Unanswered Questions in Audience Research'. *The Communication Review* 9: 101–21.

National Statistics and Coordination Board. 2012. National Statistics Office Conducts the 2012 Family Income and Expenditure Survey. *National Statistics Coordination Board*, July 23. Online: http://www.nscb.gov.ph/pressreleases/2012/PR-201207_PP1_08_fies.asp (accessed 1 April 2014).

_____. 2013. 'Despite rise in the number of families, extreme poverty among families remains steady at 1.6 million in 2012 – NSCB'. December 9. Online: http://www.nscb. gov.ph/pressreleases/2013/NSCB-PR-20131213_povertypress.asp#sthash.g8btzxUy. dpuf (accessed 1 April 2013).

National Statistics Office. 2013. Filipino Families in the Poorest Decile Earn Six Thousand Pesos Monthly, on Average in 2012. Online: http://census.gov.ph/content/filipino-families-poorest-decile-earn-six-thousand-pesos-monthly-average-2012-results-2012 (accessed 5 June 2013).

Neale, S. 1987. *Genre*. London: BFI.

Nietzsche, F. 1998[1889]. *Twilight of the Idols*. Translated by D. Large. Oxford: Oxford World's Classics.

Norgaard, K. M. 2006. '"People Want to Protect Themselves a Little Bit": Emotions, Denial, and Social Movement Nonparticipation'. *Sociological Inquiry* 76, no. 3: 372–96.

Nowicka, M. and M. Rovisco. 2009. 'Introduction: Making Sense of Cosmopolitanism'. In *Cosmopolitanism in Practice*, edited by M. Nowicka and M. Rovisco. Cornwall: Aghgate Publishing Limited.

O'Brien, N. 1993. *Island of Tears, Island of Hope: Living the Gospel in a Revolutionary Situation*. New York: Orbis.

Onega, S. 2010. *Ethics and Trauma in Contemporary British Fiction*. Amsterdam & New York: Rodopi.

O'Neill, O. 1996. *Towards Justice and Virtue*. Cambridge: Cambridge University Press.

Ong, J. C. 2009a. 'The Cosmopolitan Continuum: Locating Cosmopolitanism in Media and Cultural Studies'. *Media, Culture & Society* 31, no. 3: 449–66.

_____. 2009b. 'Watching the Nation, Singing the Nation: London-Based Filipino Migrants' Identity Constructions in News and Karaoke'. *Communication, Culture and Critique* 2, no. 2: 160–81.

_____. 2010. 'Essay on the Manila Bus Tragedy: The Safety in the Cliché'. GMA News Online, 30 August. Online: http://www.gmanews.tv/story/199737/essay-on-the-manila-bus-tragedy-the-safety-in-the-cliche (accessed 30 August 2010).

Ong, J. C. and J. Cabañes. 2011 'Engaged, but Not Immersed: Tracking the Mediated Public Connection of Filipino Elite Migrants in London'. *South East Asia Research* 19, no. 2.

_____. 2015a. 'The Television of Intervention: Mediating Patron-Client Ties in the Philippines'. In *Television Histories of Asia*, edited by Tay, J., Turner, G., and K. Iwabuchi. London & New York: Routledge.

_____. 2015b. 'Communications in Disaster Recovery: Why Facebook Matters for Yolanda Survivors But Not for the Reasons You Think.' Plenary Lecture at the Philippine Sociological Society Annual Conference. Mindanao State University, 16-18 October 2015. Online: http://philippinesociology.com/wp-content/uploads/2014/08/JCO-PSS-plenary.pdf

Ongpin, M. I. 2006. 'Holdups and Unsolved Crimes'. *The Manila Times*, 29 September, A5.

Orgad, S. 2008. 'Agency and Distance in the Representation of Suffering: A Study of UK Newspaper Coverage of the South Asia Earthquake and 7/7 London Bombings'. Paper presented at the ICA Annual Conference, Montreal, Canada.

_____. 2009. 'The Survivor in Contemporary Culture and Public Discourse: A Genealogy'. *The Communication Review* 12: 132–61.

_____. 2011. 'Proper Distance from Ourselves: The Potential for Estrangement in the Mediapolis'. *International Journal of Cultural Studies* 14: 401–21.

Orgad, S. and B. Seu. 2014. 'The Mediation of Humanitarianism: Towards a Research Framework'. *Communication, Culture and Critique*, 7 no. 1:6–36.

Ouellette, L. 2004. 'Take Responsibility for Yourself! *Judge Judy* and the Neoliberal Citizen'. In *Reality TV: Remaking Television Culture*, edited by S. Murray and L. Ouellette. New York & London: New York University Press.

Ouellette, L. and J. Hay. 2008. *Better Living through Reality TV: Television and Post-Welfare Citizenship*. London: Blackwell.

Pabico, A. P. 1999. 'Long Live the Webolution!'. In *From Loren to Marimar: The Philippine Media in the 1990s*, edited by S. Coronel. Quezon City: Philippine Center for Investigative Journalism.

Pandya, R. 2011. 'Suffering in Silence: Emmanuel Levinas and Jean-Luc Marion on Suffering, Understanding and Language'. In *Making Sense of Suffering*, edited by Hogue, B. and A. Sugiyama. The Inter-Disciplinary Press. Online: http://www.inter-disciplinary.net/publishing/id-press/ebooks/makingsense-of-suffering (accessed 2 May 2011).

Parreñas, R. S. 2001. *Servants of Globalization: Women, Migration and Domestic Work*. Stanford, CA: Stanford University Press.

Paul's Alarm. 2013. 'Presence Matters'. *Paul's Alarm,* January. Online: http://paulsalarm. wordpress.com/ (accessed 1 April 2014).

Pertierra, A. 2010. 'Consuming Modern Mexico: Television and Consumer Culture on the Mexico Belize Border'. Paper presented at the Crossroads in Cultural Studies Annual Conference, Lingnan University, Hong Kong.

Peters, J. D. 1999. *Speaking Into the Air: A History of the Idea of Communication.* Chicago: University of Chicago Press.

_____. 2001. 'Witnessing'. *Media, Culture and Society* 23: 707–23.

_____. 2005. *Courting the Abyss: Free Speech and the Liberal Tradition.* Chicago: University of Chicago Press.

Phillips, A. 1999. *Which Equalities Matter?* Cambridge: Polity.

Philo, G. 1990. *Seeing and Believing: The Influence of Television.* London: Routledge.

Pinches, M. 1987. 'People Power and the Urban Poor: The Politics of Unity and Division in Manila after Marcos'. In *The Philippines under Aquino,* edited by P. Kanks. Canberra: Australian Development Studies Network.

_____. 1999. 'Entrepreneurship, Consumption, Ethnicity and National Identity in the Making of the Philippines' New Rich'. In *Culture and Privilege in Capitalist Asia,* edited by M. Pinches. London: Routledge.

Pinchevski, A. 2005. *By Way of Interruption: Levinas and the Ethics of Communication.* Pittsburgh, PA: Duquesne University Press.

Pineda-Ofreneo. R. 1986. *The Manipulated Press: A History of Philippine Journalism Since 1945,* 2nd ed. Metro Manila: Solar Publishing.

Pingol, A. 2001. *Remaking Masculinities: Identity, Power and Gender Dynamics in Families with Migrant Mothers and Househusbands.* Quezon City: UP Center for Women's Studies.

Press, A. 1999. *Speaking of Abortion: Television and Authority in the Lives of Women.* Chicago: Chicago University Press.

Putnam, R. 2000. *Bowling Alone: The Collapse and Revival of American Community.* New York, NY: Simon & Schuster.

Radway, J. 1985. *Reading the Romance: Women, Patriarchy and Popular Literature.* Chapel Hill, NC: University of North Carolina Press.

Rafael, V. L. 1988. *Contracting Colonialism: Translation and Christian Conversion in Tagalog Society under Early Spanish Rule.* Quezon City: Ateneo de Manila University Press.

_____. 2000. *White Love and other Events in Filipino History.* Quezon City: Ateneo de Manila University Press.

Ramos-Aquino, M. 2013. '"Yolanda" Dividing Filipinos; No Thanks to Negative People'. *Manila Times,* November 15. Online: http://www.manilatimes.net/yolanda-dividing-filipinos-no-thanks-to-negative-people/53204/ (accessed 1 April 2014).

Rimban, L. 1999. 'The Empire Strikes Back'. In *From Loren to Marimar: Philippine Media in the 1990s,* edited by S. Coronel. Quezon City: Philippine Center for Investigative Journalism.

Roces, A. and G. Roces. 1992. *Culture Shock! Philippines: A Survival Guide to Customs and Etiquette.* Metro Manila: Graphic Arts Center Publishing.

Rodrigo, R. 2006. *Kapitan: Geny Lopez and the Making of ABS-CBN.* Pasig City: ABS-CBN Publications.

Romualdez, E. 1999. 'Interview: Eugenio Gabriel Lopez III'. In *From Loren to Marimar: Philippine Media in the 1990s,* edited by S. Coronel. Quezon City: Philippine Center for Investigative Journalism.

Ropeta, P. 2011. 'British Documentary Exposes Ugly Side of Manila'. ABS-CBN News Online, 27 February. Online: http://www.abs-cbnnews.com/global-filipino/02/26/11/british-documentary-exposesugly-side-manila (accessed 5 May 2011).

Rosenberg, D. A. 1974. 'Civil Liberties and the Mass Media under Martial Law in the Philippines'. *Pacific Affairs* 47. no. 4: 472–84.

Sainath, P. 2009. 'No Issues: A Recession of the Intellect'. *The Hindu*, 20 April. Online: http://www.hindu.com/2009/04/20/stories/2009042051620800.htm (accessed 10 December 2010).

Santiago, K. S. 2009. 'The Good, the Sad, the Ugly'. *Radikal Chick*, 21 December. Online: http://radikalchick.com/the-good-the-sad-the-ugly (accessed 10 December 2010).

_____. 2010. 'Marian's Freedoms'. GMA News Online, 11 September. Online: http://www.gmanews.tv/story/205510/marian-riveras-freedoms (accessed 10 December 2010).

Santos, T. 2007. '*Wowowee* Victims March to Claim "Unfulfilled" Pledges'. *Inquirer.net*, 23 January. Online: http://www.inquirer.net/specialfeatures/ultrastampede/view.php?db=1&article=20070123-45018 (accessed 10 December 2010).

Sayer, A. 2005. *The Moral Significance of Class*. Cambridge: Cambridge University Press.

Scheler, M. 1970. *The Nature of Sympathy*. Translated by P. Heath. New York, NY: Archon Books.

Schopenhauer, A. 2007[1840] *On the Basis of Morality*. Dover: Dover Philosophical Classics.

Scott, J. C. 1985. *Weapons of the Weak: Everyday Forms of Peasant Resistance*. New Haven & London: Yale University Press.

Scott, W. H. 1994. *Barangay: Sixteenth Century Philippine Culture and Society*. Quezon City: Ateneo de Manila University Press.

Sennett, R. and J. Cobb. 1973. *The Hidden Injuries of Class*. New York, NY: Vintage Books.

Seu, B. 2003. '"Your Stomach Makes You Feel that You Don't Want to Know Anything About It": Desensitization, Defense Mechanisms and Rhetoric in Response to Human Rights Abuses'. *Journal of Human Rights* 2, no. 2: 183–96.

Seu, B. and S. Orgad. 2010. 'Mediated Humanitarian Knowledge: Audiences' Reactions and Moral Actions'. Proposal to Leverhulme Trust.

Severino, H. 2006. 'Tragic Crowd'. GMA News Online, 4 February. Online: http://blogs.gmanews.tv/sidetrip/blog/index.php?/archives/54-Tragic-crowd.html (accessed 10 December 2010).

Silverstone, R. 1994. *Television and Everyday Life*. London: Routledge.

_____. 1999. *Why Study the Media?* London: Sage.

_____. 2002. 'Proper Distance: Towards an Ethics for Cyberspace'. In *Innovations*, edited by G. Liestol, A. Morrison, and T. Rasmussen. Cambridge, MA: MIT Press.

_____. 2005. 'The Sociology of Mediation and Communication'. In *The SAGE Handbook of Sociology*, edited by C. Calhoun, C. Rojek and B. Turner. London: Sage.

_____. 2007. *Media and Morality: On the Rise of the Mediapolis*. Cambridge: Polity.

Singer, P. 1973. 'Famine, Affluence and Morality'. *Philosophy and Public Affairs* 1: 229–43.

_____. 2009. *The Life You Can Save: Acting Now to End World Poverty*. London: Random House.

Sioson-San Juan, T. 1999. 'The Birth of a Medium'. In *Pinoy Television: The Story of ABS-CBN – The Medium of Our Lives*, edited by T. Sioson-San Juan. Pasig City: ABS-CBN Publishing.

Sir Martin Year. 2006. 'The *Wowowee* Effect and the Damnation of Philippine Society'. *Sir Martin Year 7*, 9 May. Online: http://sirmartin.wordpress.com/2007/05/09/the-wowowee-effect-and-the-damnation-of-philippinesociety (accessed 10 December 2010).

Skeggs, B. 1997. *Formations of Class and Gender: Becoming Respectable*. London: Sage.

_____. 2004. *Class, Self and Culture*. London: Routledge.

_____. 2006. 'Respectability and Resistance: Interview with Professor Beverley Skeggs'. *Redemption Blues*, 28 June. Online: http://www.redemptionblues.com/?p=215 (accessed 10 December 2010).

Skeggs, B., H. Wood and N. Thumim. 2009. '"Oh Goodness, I *Am* Watching Reality TV": How Methods Make Class in Audience Research'. *European Journal of Cultural Studies* 11, no. 1: 5–24.

Slote, P. 2007. *The Ethics of Care and Empathy*. London: Routledge.

Smith, D. M. 1998. 'How Far Should We Care? On the Spatial Scope of Beneficence'. *Progress in Human Geography* 22, no. 1: 15–38.

_____. 2000. *Democracy and the Philippine Media, 1983-1993*. Lewiston, NY: Edwin Mellen Press.

Social Weather Stations. 2014. 'Fourth Quarter 2013 Social Weather Survey: Adult joblessness at 27.5 %; 9% lost their jobs involuntarily, 14% resigned'. Online: http://www.sws.org.ph/pr20140212.htm (accessed 1 April 2014).

Sontag, S. 2003. *Regarding the Pain of Others*. London: Penguin.

Spivak, G. 1988. 'Can the Subaltern Speak?'. In *Marxism and the Interpretation of Culture*, edited by Nelson, C. and L. Grossberg. Urbana: University of Illinois Press.

Strathern, M. 1996. 'Cutting the Network'. *The Journal of the Royal Anthropological Institute* 2, no. 3: 517–35.

_____. 1999. *Property, Substance and Effect: Anthropological Essays on Persons and Things*. London: Anthlone Press.

Stuart Santiago, A. 2011. 'Social Media as Mosquito Press'. *Stuart Santiago*, 7 March. Online: http://www.stuartsantiago.com/social-media-as-mosquito-press (accessed 10 May 2011).

Sturken, M. 2011. 'Comfort, Irony, and Trivialization: The Mediation of Torture'. *International Journal of Cultural Studies* 14: 423–40.

Sykes, K. 2009. 'Adopting an Obligation: Moral Reasoning about Bougainvillean Children's Access to Social Services in New Ireland'. In *The Anthropology of Moralities*, edited by M. Heintz. New York: Berghahn Books.

Sznaider, N. 1998. 'A Sociology of Compassion: A Study in the Sociology of Morals'. *Cultural Values* 2, no. 1: 117–39.

Tadiar, N. 1996. 'Manila's Assaults'. *Polygraph* 8: 9–20.

_____. 2004. *Fantasy Production: Sexual Economies and Other Philippine Consequences in the New World Order*. Quezon City: Ateneo de Manila Press.

Tait, S. 2011. 'Bearing Witness, Journalism and Moral Responsibility'. *Media, Culture and Society* 33: 1220–35.

Tan, M. 2006. 'Dare Imagine'. *Inquirer.net*, 10 February. Online: http://www.inquirer.net/specialfeatures/ultrastampede/view.php?db=0&article=20060210-65663 (accessed 10 December 2010).

Tester, K. 2001. *Compassion, Morality and the Media*. Buckingham: Open University Press.

Thompson, E. P. 1978. *The Poverty of Theory and Other Essays*. London: Merlin.

Thompson, J. 1990. *Ideology and Modern Culture*. Cambridge: Cambridge University Press.

_____. 1995. *The Media and Modernity: A Social Theory of the Media*. Cambridge: Cambridge University Press.

Time Magazine. (2014). 'The Selfiest Cities in the World'. Online. http://time.com/selfies-cities-world-rankings/

Tiongson, N. 1994. 'Philippine Theater'. In *CCP Encyclopedia of Philippine Art*, vol VII. (1st ed.), edited by N. Tiongson. Manila: Cultural Center of the Philippines.

Tolentino, R. 2001. *National/Transnational: Subject Formation and Media in and on the Philippines*. Quezon City: Ateneo de Manila University Press.

_____. 2007. 'Kulturang Popular at Pakiwaring Gitnang Uri'. *Rolandotolentino*, 12 November. Online: http://rolandotolentino.blogspot.com/2007/11/kulturang-popular -at-pakiwaring-gitnang.html (accessed 10 December 2010).

_____. 2010. 'Journalists and Media Workers Should Know Crisis Reporting, Aquino Administration Must Be Held Accountable for Disorganized Police'. *Rising Sun*, 27 August. Online: http:// risingsun.dannyarao.com/2010/08/27/journalists-and-media-workers-should-know-crisisreporting-aquino-administration-must-be-held-accountable-for-disorganized-police (accessed 10 December 2010).

_____. 2011. 'Kabataang Katawan, Mall, at Syudad: Gitnang Uring Karanasan at Neoliberalismo'. Paper presented at the Space, Empire, and the Postcolonial Imagination Conference, Ateneo de Manila University, Quezon City, Philippines.

Tomlinson, J. 1999. *Globalization and Culture*. Cambridge: Polity.

Torres, J. F. G. T. 2006. '"Treated Like Animals" – Lead Investigator'. Inquirer.net, 7 February. Online: http://www.inquirer.net/specialfeatures/ultrastampede/view.php?db=0&article=20060207-65404 (accessed 10 December 2010).

Tronto, J. C. 1993. *Moral Boundaries: A Political Argument for an Ethic of Care*. London: Routledge.

Uy, I. 2009. 'What Does It Mean to Be *Jologs*? (Or My Violent Reaction to SNSD's Popularity in the Philippines)', *Sankofa*, 27 October. Online: http://blog.ianuy.com/2009/10/27/what-does-it-mean-to-be-jologs-or-my-violent-reaction-to-snsds-popularity-in-the-philippines (accessed 10 December 2010).

Van Zoonen, L. 2003. 'After *Dallas* and *Dynasty* We Have... Democracy'. In *Media and the Restyling of Politics*, edited by J. Corner and D. Pels. London: Sage.

Venture for Fundraising. 2003. *2003 Nationwide Survey on Giving*. Quezon City: Venture for Fundraising.

Vestergaard, A. 2009. 'Identity and Appeal in the Humanitarian Brand'. In *Media, Organisation and Identity*, edited by L. Chouliaraki and M. Morsing. London: Palgrave Macmillan.

Virola, R., Encarnacion, J., Balamban, B., Addawe, M. and M. Viernes. 2013. 'Will the Recent Robust Economic Growth Create a Bourgeoning Middle Class in the Philippines?' Paper presented at the 12th National Convention on Statistics, EDSA-Shangri-La Hotel, Mandaluyong City.

White, M. 2006. 'Investigation *Cheaters*'. *The Communication Review* 9: 221–40.

Wood, H. and B. Skeggs. 2009. 'Spectacular Morality'. In *The Media and Social Theory*, edited by D. Hesmondhalgh and J. Toynbee. London & New York: Routledge.

Yapan, A. 2009. 'Nang Mauso ang Pagpapantasya: Isang Pag-Aaral sa Estado ng Kababalaghan sa Telebisyon'. *Plaridel*, 11.

Young, A. 1997. 'Suffering and the Origins of Traumatic Memory'. In *Social Suffering*, edited by A. Kleinman, V. Das, and M. Lock. Berkeley: University of California Press.

Zelizer, B. 2008. 'Journalism Ethics from the Bottom Up'. Paper presented at Ethics of Media Conference. CRASSH, University of Cambridge, Cambridge, UK.

Zigon, J. 2007. 'Moral Breakdown and the Ethical Demand: A Theoretical Framework for an Anthropology of Moralities'. *Anthropological Theory* 7, no. 2: 131–50.

INDEX

ABS-CBN 2, 8, 54–7, 66, 68, 71, 74, 84–5, 90, 92–6, 118, 122–3, 126, 133, 140–3, 145, 151, 172, 184, 186, 188n1, 188n4, 189n5, 190n1, 191n4, 191n5

age 1, 9, 25, 42, 50, 62, 64, 89, 100, 107, 159, 165, 168, 179

agency: in the representation of suffering 12, 44, 46–8, 50, 90, 107–18, 133–4, 140, 150, 155, 166–8, 172, 184, 192; in media participation 12, 90, 95, 107–18, 166–8; of the poor and working-class 29, 33, 53, 107–18, 128; texts as enacting 18

Aguilar, Filomeno 6, 28–9, 62–3, 188n1, 191n8, 192n9

anthropology of moralities 9–10, 15, 19–24, 31, 34, 36, 151, 153, 155, 166, 177–8, 181–2, 185, 187n1

appropriation 78–81, 136, 177

Arendt, Hannah 17, 27, 47, 82, 171

Aristotle 27

attention 4, 13, 24, 46, 48, 65, 70, 109–11, 116, 121, 126–32, 134–5, 137, 147–8, 150–2, 154, 175, 192n12

audience studies 5, 45, 48–52, 119, 172

audience-centred perspective (including bottom-up approach and anthropological ethics of media) 1, 5, 11, 13, 39, 48, 59, 153, 155, 166–9, 185

audiences: avoidance strategies 5, 7, 26, 28–32, 49–51, 61, 72–7, 87, 128–9, 148–9, 155–8 (*see also* denial);

moral duties, obligations and responsibility, especially to sufferers 1, 3–5, 7, 10–3, 16–8, 20, 25–30, 36, 40, 42–3, 46, 49–51, 61, 72, 77, 115–6, 120–1, 126, 129, 132–3, 135–8, 158, 160–3, 166, 169, 178, 187n2

authenticity 13–4, 23, 41–2, 81–3, 86, 90, 95, 101, 105–7, 112, 115, 120, 134, 138–9, 157, 162–4

Bankoff, Greg 4, 6, 30, 34–6, 63–4, 86, 121, 131, 142, 150, 158

Big Brother 73–4, 104, 189n9, 190n10

the body 33, 76, 80–1, 101, 107, 163, 191n8

Boltanski, Luc 3–5, 7, 24–5, 49, 61, 86, 110, 120–1, 150, 161, 171

Born, Georgina 15, 19, 20, 52, 166, 169

Bourdieu, Pierre 33, 36, 51, 64, 108, 139, 166

Cabañes, Jason 7, 30, 129, 150, 159, 190

Cannell, Fenella 4, 6, 28–9, 35–6, 63–4, 77, 86, 112, 158, 170, 192n12

Catholicism 6, 28, 31, 64, 79–80, 117, 138, 140, 150, 159, 179, 183, 188n3

chapter summaries 9–14

charity: appeal 111, 120, 126, 130, 139–142, 151, 161; as concept 25, 26, 37, 118; donating to 61, 87, 97, 120, 137–8, 150, 160, 171, 174; television network charities 13, 56–57, 84, 86, 125–126, 139–43, 151, 154; volunteering for 71–2, 79–80, 83, 150, 179, 183–4

Chouliaraki, Lilie 1, 3, 4, 7, 9–10, 16, 18–20, 24, 35, 39, 44–8, 53, 62, 99, 121, 129–33, 137, 140, 150–1, 154, 161, 164, 169, 178

Couldry, Nick 3, 16, 18, 39–40, 43–4, 52,
62, 72, 83–7, 96, 109–10, 117, 119,
122, 158, 159, 163, 165, 168, 169,
171–2, 187n1
class: demographic descriptions in the
Philippines 2, 7, 55–6, 62–4; 188–9;
divides 2, 8, 14, 29, 36, 169–73;
moralities 12, 24, 30, 33–4, 51,
162; as social category 1–3, 9–10,
18, 23, 37
compassion: as concept 4, 7, 9, 15, 25–8,
30, 33; discourses of 5, 8–9, 44,
45, 47, 48–51, 89–90, 96, 99–100,
105, 109–10, 112–8, 120, 122, 128,
134–5, 138, 140, 145, 147–51, 159,
161, 167–71, 178, 182–3, 185;
fatigue 1, 5, 11–2, 14, 44, 46, 48–51,
119, 128, 154–5, 157
confessional 9, 90–2, 101, 103, 108, 112
coping mechanisms (including endurance
of suffering) 7, 10, 35, 41, 59, 63,
117, 121, 138, 151, 169; sociology of
25, 34, 36, 158, 187n1
cosmopolitanism 18–20, 135, 138, 150,
159–61, 166, 175
culture of disaster 7, 34–5, 121, 142,
158, 174

Das, Veena 4, 32–44, 111, 116, 156
David, Randy 6–7, 25, 28, 34, 36, 56, 62,
117, 123, 157, 169, 187n1
Dayan, Daniel 17–8, 24, 187n2
De Quiros 2, 57, 85, 94, 139, 142, 145,
151, 165, 171
de-Westernizing approach 4, 18, 24, 34,
132, 150, 155–8
dehumanization 4, 174
denial 3–4, 8, 30, 49–51, 56, 69, 72, 76,
81, 126, 153, 158, 160, 164, 175
(*see also* distancing strategies)
deservedness (or deserving) 87, 90, 95, 99,
100, 101, 105–6, 110, 112, 114, 118,
120, 128, 135, 139, 144, 148, 157,
162–3, 170
dignity 12, 27, 46, 92, 110, 140, 150, 159,
163
direct experiences with media 3, 8–9, 44,
52–4, 62, 83–6, 122, 168, 180, 191

distancing strategies 30, 41, 49–51, 72,
76–7, 132–8 (*see also* audiences,
avoidance strategies *and* denial)

elite (or the rich) 58, 67; consumption of
media 64–72, 129–30; judgments of
media 8, 53, 72–81, 90–6, 109–18,
128, 131–8, 139–42; judgments of
the poor 95 (*see also jejemon, jologs*);
moralities, 12, 81, 138, 174
Ellis, John 4, 41–2, 78, 162
ethics: audience 11, 40, 44, 48–52, 59,
86–7, 89, 113, 168–9; codes of 12,
16, 187; ecological 11, 40, 44–5,
52–4, 59, 62, 83, 86, 90, 113, 116,
122, 164, 167, 169, 185; situated 9,
18–20; textual 11–2, 40, 44–9, 52–3,
89, 107, 133, 138, 150, 157, 164,
167,169, 185
ethnicity 1, 3, 18, 42, 44, 50, 100, 107,
132–8, 158, 165, 191n8
ethnography 1, 9, 14, 19, 20, 24, 29, 31,
33–4, 40, 43, 50, 59, 122, 137, 150,
154–7, 178, 181, 185
everyday life 21, 24, 32, 59, 67, 69, 72, 76,
85, 86, 90, 112, 116, 119, 121, 141,
153, 161
exploitation 4, 12, 51, 53, 87, 90, 95–6, 99,
109–13, 139–40, 150–1, 167–8, 172,
182, 185

Facebook 66, 82, 129, 137, 143, 145, 147,
189n3, 190n15
first principles 16–21, 166
focus group 119, 181
Friendster 66, 189n3
Frosh, Paul 4, 13, 41–2

game show 2, 13, 90–118
gender 9, 12, 25, 31, 42, 50–2, 54, 62, 64,
73, 77, 81, 89, 128, 159, 165, 178
GMA Network 54–8, 65, 68, 71–3, 79–80,
85, 94–5, 122–6, 143, 145, 184, 186,
189n5, 191n4, 192n10

Haiyan, Typhoon 58, 173–4
Hall, Stuart 43
Hill, Annette 106

Höijer, Birgitta 5, 49–50, 89, 99–100, 105, 110, 119, 128, 132, 134, 179, 182
hospitality 7, 52, 84, 87, 11, 156, 159, 168–9

identification 5, 16, 19, 25, 27, 46, 51, 68, 79, 92, 105, 109, 128, 135, 157, 172, 180
India 2, 8, 32, 56, 69, 111, 116, 126, 175

jejemon 82, 173, 190n15
Johnson, Mark 6–7, 28–30, 158, 163
jologs 74–83, 85–7, 109, 136, 160, 162–5, 190n15
journalist 2, 8, 52, 57, 63, 68, 76, 78, 83–5, 123–8, 132, 138, 143–5, 189, 191n4

Kerkvliet, Benedict 6–7, 29–30, 51, 63, 77, 86, 119, 158, 171
Kyriakidou, Maria 5, 10, 51, 89, 119, 132, 137, 156, 179

Laidlaw, James 3, 21, 23–4, 163
lay moralities or normativities 9, 15, 20–3, 33–4, 39, 53, 59, 116, 120, 139, 166–7, 177–8, 187–8; lay media moralities 51, 90, 113, 120, 139, 147, 150–1, 164–7
Levinas, Emmanuel 10, 17, 20, 45
Livingstone, Sonia 10, 39, 43, 49, 120, 129, 147, 155, 169
lower-class: consumption of media 64–72, 126–9; judgment of media 81–3, 109–18 126–9, 132–8, 139–42

Madianou, Mirca 3, 29, 39, 44, 52–3, 66–7, 72, 83, 122, 147, 155, 161, 168–9, 171, 178, 189n3, 191n7
Manila 30, 36, 65, 67, 80, 123, 126, 133, 143, 150, 173, 175
Marcos, Ferdinand 6, 54, 123
martial law 6, 54, 56, 123, 173
the *masa* (masses) 2, 8, 32, 56, 58, 63, 71, 73–5, 79–82, 109, 112, 116, 124, 129, 156, 160, 162–3, 166, 167, 170, 173, 185
McKay, Deirdre 6, 28–9, 51, 107, 137–8, 150, 160

media: international 11, 83, 166; pilgrimage 83–6, 93–6, 110–3; power 13, 43–4, 165, 168–72; social critique of 12, 14, 164; systems 10; talk 3
media studies, moral turn in 1, 3, 16–20
mediapolis 17–8
mediated centre 62, 83–7, 95–6, 111–3, 118, 155, 162–8, 170–1, 191n8
mediation 6, 10, 11, 13–4, 16, 19, 37, 39–0, 43–5, 49, 51–3, 59, 61, 77, 86, 96, 118, 142, 146, 153–5, 158, 167–75, 178, 180–1, 187n1
mediatization 44
methodology 177–86
middle-class: consumption of media 64–72, 131–2; judgment of media 72–81, 109–18, 131–8, 139–42; migrants 29; moralities 138; neighborhoods 11
Miller, Daniel 29, 66–7, 74, 189n3
Miller, David 21, 24
Moeller, Susan 4–5, 24, 46, 48, 52, 89, 132, 155, 157
Monzon-Palma, Tina 141
moral justification 12–3, 22, 26, 50–1, 64, 87, 120, 132–8, 157, 161, 163, 182

news 4, 11, 13, 18–9, 25, 41, 46–50, 56–7, 67–73, 82–6, 119–51, 155, 157, 161, 165, 168, 170, 174, 180, 182–5

Ondoy, Typhoon 2, 13, 57, 142–7, 180, 184
ontological security 70, 121, 130–1, 146, 147, 157
Orgad, Shani 4–5, 10, 44, 46–8, 89, 116–8, 132, 141, 158, 166, 174
the other 1, 3–4, 10, 16–8, 40, 45–7, 72, 76, 78, 89, 121, 139, 159, 162
over-representation 2–3, 14, 61, 76, 90, 92, 118, 157, 159, 165–8, 171, 173

patron-client ties 30, 53, 57, 118, 171
Peters, John Durham 4, 13, 16–7, 20, 40–2, 44–5

Philippines: government 136; history
    of 6, 44; media conventions in 12,
    44, 54–6, 64, 84–6, 93–6, 120,
    122–6; problem of witnessing
    suffering in 28–30; sachet
    economics in 66, 174; as 'social
    networking capital of the world'
    66; as 'texting capital of the world'
    66, 175; weak state 12–3, 84, 120,
    122–6, 154
philosophical approach to media ethics 5,
    9, 20
phronesis 9, 18, 178
Pinches, Michael 28–9, 36, 62–4, 75, 81,
    100, 134, 158
Pinchevski, Amit 4, 13, 17, 20, 40–2, 45
pity: as emotional response 5, 7, 27, 45, 49,
    73, 80, 99, 109, 117, 149, 171, 183,
    185; idioms of 4, 35, 111–3
the poor: for their practices, *see* coping
    mechanisms, patron-client ties,
    resilience, resistance; for strategies
    of ignoring, *see* denial, distancing
    strategies
poverty: glorifying 92; poverty index 7, 63;
    sanitizing 2, 4, 83
Poverty of Television (an explanation of) 2, 3
prayer 64, 138
proper distance 14, 46–7, 116, 126

reality TV 4, 33, 47, 53, 62, 90, 106, 117,
    132, 162, 173
recognition 16, 33–5, 53, 62–3, 82–3,
    86–7, 110–3, 116–8, 138, 157, 159,
    161, 164, 167, 168, 170–1, 173–4,
    185
recruitment of participants in media 53,
    179
redistribution 28, 34, 62, 86, 112–3, 158,
    165, 168, 171
religion 3, 6, 9, 18, 25, 31, 50, 62, 64, 138,
    151, 159, 178–9
resilience 6–7, 132, 147, 151, 174
respectability 12, 23, 30, 33, 35, 53, 62, 75,
    85, 90, 95, 109, 114, 139, 141, 147,
    150, 159, 162–5, 184
Revillame, Willie 92, 94–5, 103, 108, 143,
    190n1

Sayer, Andrew 33–4, 62, 87, 163
Scott, James 6, 32
Seu, Bruna 3–5, 10, 47, 49–51, 61, 72, 76,
    81, 121, 135–6, 156, 158, 164, 166
shock effect 7, 45, 130, 175
Sichuan earthquake 13, 132–8, 150, 160,
    181–2
Silverstone, Roger 1, 4–5, 10, 16–20, 23,
    25, 27–8, 39–40, 44–7, 52, 61, 70,
    72, 76, 116, 121, 123, 147, 155, 157,
    159–3, 166, 169–71, 177–8, 187n2
Skeggs, Beverly 4, 24, 33, 51–3, 62, 64,
    76, 87, 90, 92, 108, 117–20, 122,
    139, 147, 158–9, 162–3, 168, 171,
    190n12
social media 66, 173–5
Sontag, Susan 3, 15
suffering: 'at its worst' 4, 46–47, 141, 174;
    coping with *see* coping mechanisms;
    definitions of 4, 24; distant 4–6, 11,
    13, 18, 25, 32, 36, 39–45, 48–54,
    132–8, 140, 143, 145, 146, 160,
    185; everyday 5, 151, 154, 156–8;
    strategic 107–13
slum communities 11, 56, 63, 66–8, 140,
    179, 180, 187n1
switching off or looking away 12, 14,
    49–50, 72–7, 89, 114, 120–2, 126,
    129–30, 142, 147, 148, 150–1, 158,
    161–2, 166, 169

Tadiar, Neferti 7, 30, 51, 61, 66, 70, 73,
    76, 116, 122, 153, 158, 170
Tait, Sue 42
television: cable television 65, 70,
    122–3, 188n4; as flow 67–70; as
    interruption 70–2; as object 64–83;
    ratings 55–6, 67, 74, 93, 95–6,
    109, 110, 122, 125–6, 139, 190n4;
    studio 84–6, 93–6, 99, 110–3, 125,
    143, 180, 191n6 (*see also* media
    pilgrimage); television (and media)
    landscape in the Philippines 10,
    54–5; viewers (*see* audiences)
text *see* media text
Tiangco, Mel 57, 125–6
Tolentino, Roland 8, 55, 62, 66, 73, 126,
    188n1

upper-class: *see* elite

victim: asserting 31–32, 35, 118–9,
    132–136; representing sufferers as 6–7,
    12, 89, 107, 114, 116, 162, 168, 185
victimhood 31–2, 111–2, 132–8 (*see also*
    agency)

Western context of witnessing suffering
    4–7, 10, 17–9, 34–5, 39, 41–3, 46,
    48, 72–3, 76, 86, 132, 150, 153,
    155–8, 170, 174, 185
witnessing 6, 10–4, 25, 36, 39–45, 50, 162;
    moral problems in 25–30
women 31–3, 69, 70, 77, 80–2, 93, 95,
    116, 180, 189n4, 191n6

Wood, Helen 4, 33, 52–3, 76, 87, 90, 92,
    117–8, 120, 122, 147, 158–9, 162–3,
    168, 171
working-class, excess 76–7, 86, 113, 130,
    139, 153
Wowowee 12–3, 67; 90–118, 127–8, 139,
    143, 163–4, 167, 168, 180, 185,
    190n1

Zelizer, Barbie 15–6, 19–20, 169, 187
zones: of danger 18, 24, 62, 72, 77, 108,
    131, 142, 156, 158–61; of safety 18,
    62, 72, 77, 86, 108, 129, 132, 136,
    142, 151, 158–61

Lightning Source UK Ltd.
Milton Keynes UK
UKOW02n1236210515

252034UK00002B/37/P